Framing Big Data

Maria Cristina Paganoni

Framing Big Data

A Linguistic and Discursive Approach

Maria Cristina Paganoni
Dipartimento di Scienze della
Mediazione Linguistica e di Studi
Interculturali
Università degli Studi di Milano
Milan, Italy

ISBN 978-3-030-16787-5   ISBN 978-3-030-16788-2   (eBook)
https://doi.org/10.1007/978-3-030-16788-2

This Palgrave Pivot imprint is published by the registered company Springer Nature Switzerland AG
The registered company address is: Gewerbestrasse 11, 6330 Cham, Switzerland

To my sons and daughter

Acknowledgements

This work sees the light within the National Research Project (PRIN 2015TJ8ZAS) 'Knowledge Dissemination across Media in English: Continuity and Change in Discourse Strategies, Ideologies, and Epistemologies', funded by MIUR (Italian Ministry of Education, University and Research).

My warmest thanks go to Ornella Bramani, head of Lubrina Publisher, who edited a shorter, preliminary version of this work, for releasing the rights for this updated and expanded volume.

I would also like to thank Professor Emeritus of Internal Medicine Claudio Rugarli, of the San Raffaele Hospital and Scientific Institute in Milan, for his insightful observations regarding machine learning in the future of medicine.

ACKNOWLEDGEMENTS

CONTENTS

LIST OF TABLES

CHAPTER 1

Big Data in Discourse

Abstract The chapter explains the meanings of big data and raises excitement and doubt over its use within the wider societal context. To introduce the critical features of big data analytics, it recapitulates the unfolding of scientific paradigms over history, with the framework of "data-driven" science currently taking shape amid contestation. It discusses how the big data debate urges insights from different disciplines to assay the epistemological, anthropological and ethical changes brought about by technology advancements. Academic conversations surrounding novel approaches to knowledge production, dissemination and research ethics are also summarised. Finally, a linguistic and discourse-analytic approach is seen as conducive to novel insights into three crucial domains, i.e., the global news media, healthcare and the recent regulation of data protection in the European Union.

Keywords Big data · Data-driven science · Epistemology · Ethics · Ontology

1.1   BIG DATA AND TECHNOLOGICAL DISCOURSE

Big data has of late been attracting great attention in business, government and science, from where it is propagating into wider societal circles that respond to new forms of knowledge and contribute to its dissemination. Aiming at a deeper understanding of what big data means and of its impact in society, this work intends to prioritise a linguistic and discursive approach to the big data debate. That phrases such as artificial intelligence, the internet of things and, lastly, big data should be coming to the fore in discourse can be ascribed to the hype cycle of technological expectations and the widespread adoption of computational and digital practices in contemporary society.[1] However, the conceptualisation and application of innovative socio-technical constructs in key societal domains like science, medicine, law, business, politics and government policy are still under-researched.

This imperfect understanding reverberates in the polarised coverage big data receives in the news media and in public and policy discourse. Scientific and technical innovation is hailed as "a major technological revolution that will reshape manufacturing industries and social and economic life more broadly" (Reischauer 2018, p. 26) but, at the same time, the advances brought about by big data tend to be associated with feelings of unease when loss of control and potential threats to human rights are exposed. Examples of this include data aggregation and access control of data repositories on the part of governments and corporate ventures in which the surveillance trends of big data are intensified (Lyon 2014), from the Snowden revelations of the US intelligence's phone surveillance program in June 2013[2] to the Cambridge Analytica data leak in March 2018.[3] Against this background which is still shaping itself, this research work sets out to offer an overview of the big data debate and a fresh look into the interweaving of discursive perspectives coming from different social actors and fields which are not limited to the technological domain and its experts. In sum, the aim is "to dismantle claims about the given and irrevocable facticity of data formats and data analytics so as to explore ways of reimagining their status and implications" (Pentzold and Fischer 2017, p. 2).

Searching the etymology of "big data" is an excellent starting point to investigate the potential of discourse and storytelling in the dissemination of technological innovation among lay publics and their contribution to the making of a contemporary technological imaginary. While *data*, as the plural of the Latin loan word *datum*, is a countable noun, "*big data* is grammatically a mass noun, a conceptual shift from single units of information to a homogeneous aggregate" (Puschmann and Burgess 2014, p. 1694). The shift explains why the noun phrase is often capitalised ("Big Data"),[4] as it is perceived as a collective entity that is conceived as a proper noun. The big data that mesmerises the public focus on science today emerged in common parlance in the early 1990s, but began "gaining wider legitimacy only around 2008" (Boellstorff 2013, n.p.).

Retracing the origin of the phrase has been compared to an "etymological detective story" (Lohr 2013, n.p.). One claim is that big data "was first used by John Mashey, retired former Chief Scientist at Silicon Graphics, to refer to handling and analysis of massive datasets" (Diebold 2012, quoted in Kitchin 2016, p. 1) around the mid-1990s. By contrast, there are others who believe that "[l]ikely the first use of 'big data' to describe a coherent problem was in a publication by Michael Cox and David Ellsworth in 1997 attributing the term to the challenge of visualizing large datasets" (Metcalf et al. 2016, p. 4). In any case, since the last decade of the twentieth century, big data has quickly turned into a catchphrase for tech industries, with IBM as the first major company that has capitalised on it (Zikopolous et al. 2012), constructing a very optimistic social reality around big data through discourse.[5] In November 2011, the American *Popular Science* magazine published a special issue entitled "Data Is Power", highlighting the empowering effects of data.

Despite several attempts to categorise the identifying features of big data,[6,7] a stable definition has still not been reached in the literature and "the lack of a systematic meta-discourse surrounding the polysemy of Big Data" (Portmess and Tower 2015, p. 4) is apparent in the debate. Nonetheless, a broad consensus exists on the fact that the "bigness" of data is not just a matter of volume but of mindset (boyd and Crawford 2012; Kitchin 2014a, b; Shin and Choi 2015), as it leads to the break-up of existing research approaches and "the capacity to search, aggregate, and cross-reference large data sets" (boyd and Crawford 2012, p. 663). It is argued, quite convincingly, that it is the complexity of the task, rather than data magnitude, that qualifies big data: "Big Data discursively refers to a qualitative shift in the meaning of data, in not just the amount of data (approaching exhaustiveness) but also its quality (approaching a dynamic, fine-grained relational richness)" (Chandler 2015, p. 836).

In actual fact, though somewhat approximately in light of big data's defining features (Kitchin, *supra* note 6), a variety of data sets are commonly included under this denomination: government registries and databases, national surveys and censuses, online multimodal repositories, corporate transactional data, health data collections and biobanks, user-generated content on social media, web indexes and directories, measurements from sensor-embedded environments, surveillance and communication metadata tracing digital footprints and so on. The network of internet-connected technological contrivances capable of collecting and exchanging data is referred to as the internet of things. The label describes a variety of electronic devices, appliances, sensor-laden objects, vehicles, smart watches, wristbands, garments and wearables, self-tracking technologies and apps that are able to interact with the real-world environment, in quite an unprecedented and revolutionary way (Paganoni 2017).[8] "Datafication" (Lycett 2013; van Dijck 2014; Chandler 2015) is the constant generation of data flows that represent people's digital footprints today and it applies to human experiences previously hardly quantifiable on such a scale, such as the mapping of "friendships" on social media, behaviour and preferences. It raises "problematic ontological and epistemological claims" in terms of "*belief* in the objective quantification and potential tracking of all kinds of human behavior and sociality through online media technologies" and "trust in the (institutional) agents that collect, interpret, and share (meta)data culled from social media, internet platforms, and other communication technologies" (van Dijk 2014, p. 197, original emphasis).

Big data has travelled from corporate storytelling and the precincts of high-tech companies to permeate "academic and popular culture descriptions [*whose*] phraseology evokes the notion of an overwhelming volume of data that must somehow be dealt with, managed and put to good use" (Lupton 2014, p. 107). The socio-technical phenomenon of big data "triggers both utopian and dystopian rhetoric" (boyd and Crawford 2012, p. 663), in which the ambivalence felt towards computer technologies finds an outlet in linguistic creativity, i.e. in the polarised sentiment of metaphorical language whereby the data revolution may turn into a destructive data deluge (van den Boomen 2014).

"It may seem marketing gold, but Big Data also carries a darker connotation, as a linguistic cousin to the likes of Big Brother, Big Oil and Big Government" (Lohr 2012, n.p.). The collective vision about big data includes social expectations and fears that are also conveyed through language: "There is a non-neutral discursive interplay between the terms we use to talk about technology, the way we subsequently understand a technology, and the way we consequently choose to act upon this" (Pots 2014, p. 3). This is the reason why framing the big data debate is ultimately a way to look into a number of impelling issues that affect not just corporations but society at large.

1.2   QUESTIONING THE BIG DATA NARRATIVE

The assumption that best captures the main controversies in the debate is that "contemporary discussions concerning big data have been technologically biased and industry-oriented, leaning toward the technical aspects of its design" (Shin and Choi 2015, p. 311). In other words, the big data debate arises from the conviction that a technology-only perspective is too confining and that big data involves social, cultural and ethical aspects that cannot be left unquestioned. Since it is reshaping traditional epistemological paradigms (Kitchin 2014a; Cowls and Schroeder 2015), big data represents a major research object also for the humanities and social sciences, where academics are challenged to conceptualise it, argue its implications for scholarship and relevant methodologies, and devise an appropriate code of ethics that will circumvent technological determinism (Halford and Savage 2017; Woodfield 2017).[9]

In order to critically tackle the big data debate with the tools of discourse studies, it is important to trace its basic outlines within a number of societal domains. At the same time, this is such a wide canvas that it would be unrealistic to aspire to map it in its entirety. This work has the more limited objective of grounding its premises in the current theorisation about big data in the social sciences and the humanities before proceeding with the linguistic and discursive analysis of the knowledge dissemination process.

1.3   EPISTEMOLOGICAL, ANTHROPOLOGICAL AND ETHICAL ASPECTS IN THE BIG DATA DEBATE

In the social sciences and the humanities, critical issues about big data are associated with the break-up of established epistemological, anthropological, and ethical paradigms that have been brought about by the so-called data revolution (Kitchin 2014b). The first perceived fault line of disruption concerning big data is *epistemological*. "Big Data has emerged as a system of knowledge that is already changing the objects of knowledge", while it is undoubted that the big data debate "reframes key questions about the constitution of knowledge, the processes of research [and] how we should engage with information" (boyd and Crawford 2012, p. 665). Against all claims of objectivity, big data first and foremost "challenges established epistemologies across the sciences, social sciences and humanities"

(Kitchin 2014a, p. 1), leading to the emergence of a fourth paradigm of science, that of *data-intensive*, expository science (Hey et al. 2009).

The "argument that computing was fundamentally transforming the practice of science" was made by Jim Gray, "a database software pioneer and a Microsoft researcher" (Markoff 2009) and a Turing Award winner, in his last talk to the Computer Science and Telecommunications Board on 11 January 2007.[10] Gray divided the evolution of science into four periods or paradigms. One thousand years ago, science was experimental in nature, describing natural phenomena, a few hundred years ago it became theoretical, using models and generalisations, a few decades ago it turned into a computational discipline, and today it is data-intensive (Hey et al. 2009). Theory, experimentation, computation and, now, data-intensive science "can be differentiated by the amount of data they produce and consume" (Nelson 2009, p. 6).

Over the last decade, the extent to which data-intensive science marks an epistemological disruption has been articulated from different positions. When Chris Anderson, editor-in-chief at *Wired* and one of the big data evangelists, announced the "end of theory" in 2008, his "provocative piece [...] was referring to the ways that computers, algorithms, and big data can potentially generate more insightful, useful, accurate, or true results than specialists or domain experts who traditionally craft carefully targeted hypotheses and research strategies" (Graham 2012, n.p.). The "end of theory" marked the onset of a new empiricism whereby big data computing is viewed as being capable of generating knowledge without the need for pre-existing models, as big data can speak for itself. The claim Anderson advanced is that the availability of huge datasets and computing power enable linear cause-effect understandings to be superseded by correlations between data. The messiness of big data, which is seen as a reflection of the complexity of its nature, implies "a broader ontological shift in the understanding of the world and its problems [...] [*a*] shift to inductive reasoning – or to letting the 'data' do the work instead of the theorist" (Chandler 2015, p. 834). In psychology, it is called *apophenia*, i.e. the perception of patterns and meaningfulness in unrelated things (boyd and Crawford 2012, p. 668).

Since 2008, however, objections to the "no theory" claim have been raised by critics who contest the "mythology" of big data, i.e. "the widespread belief that large data sets offer a higher form of intelligence and knowledge that can generate insights that were previously impossible, with the aura of truth, objectivity and accuracy" (boyd and Crawford 2012, p. 663), stressing the need for theoretical models also in big data and machine learning (Frické 2015; Mazzocchi 2015). The empiricist epistemology is contested by proponents of "data-driven" science, whose paradigm still sees theory as being useful to knowledge discovery (Kitchin 2014a), arguing that "[b]ig data need big theory too" (Coveney et al. 2016, p. 1). This point is well illustrated in an interview to Rob Kitchin on the LSE Impact of Social Sciences Blog, run by the London School of Economics and Political Science, in which the social scientist responded to Chris Anderson's proclamations:

> This is a problematic position for a number of reasons. Big data, despite its attempts to be exhaustive is always partial, with gaps, biases and uncertainties. Moreover, the data do not come from nowhere, but are produced by systems that are designed and tested within scientific frameworks, and are surrounded by an assemblage of different contexts and interests. The production of big data is thus driven by knowledge and other factors which mean the data never simply speak for themselves. (Williams 2014, n.p.)

Another important point which is made in favour of data-driven versus data-intensive science is that "our semantic conceptions of meaningful information preclude the possibility of proto-epistemic data, or pure, unadulterated, unstructured data" (Portmess and Tower 2015, p. 4), i.e. it is impossible to separate the technologically informed contexts in which data is inscribed from "human intentionalities" (Verbeek 2011, p. 58).

In retrospect, besides, the "end of theory" stance is now regarded as a provocation and an overstatement thanks to Chris Anderson himself, who backtracked from his claims soon after the publication of the article (Cowls and Schroeder 2015, p. 456). In actual fact, big data may more easily test or refine theory than in the past if big data analytics does not succumb to "hornswoggling", i.e. "unsound statistical fiddling" (Frické 2015, p. 651). The need for interpretation is also the point made by Pentzold and Fischer (2017), together with a number of critical data studies that question the ontology and epistemology of data (Bowker and Star 1999; Boellstorff 2013; Andrejevic 2014; Crawford et al. 2014; Halavais 2015) and speak of reflexive knowledge rather than causal knowledge. "In essence, the imaginary of Big Data resolves the essential problem of modernity and modernist epistemologies, the problem of unintended consequences or side-effects caused by unknown causation, through work on the datafication of the self in its relational embeddedness"

(Chandler 2015, p. 11). What this argument implies is that recording data flows from the self and the world makes it possible to collect personal information and social practices on an unprecedented scale at the micro level of analysis. At that point, big data analytics can interpret their ontological complexity, i.e. not "by generic laws and rules but by feedback loops and changes through iterative and complex relational processes" (Chandler 2015, p. 18). In medicine, for example, "machine-learning algorithms can improve the accuracy of prediction over the use of conventional regression models by capturing complex, nonlinear relationships in the data" (Chen and Asch 2017, p. 2507).

Anthropological. The radical epistemological and ontological claims made on behalf of big data are related to "the rubric of posthumanism" (Chandler 2015, p. 3) and "the constitution of new subjects" (Fisher 2010, p. 231) that are concomitant with the demise of an anthropocentric worldview. In the Petabyte Age, the widespread use of sensors, machine learning and algorithmic regulation seems to have heralded a posthuman world in which humans will integrate with machines and machine intelligence. In sum, a major consequence is that "Big Data approaches discursively turn posthumanism into a mainstream research agenda" (Chandler 2015, p. 18), and not just in life sciences, but in the humanities and social sciences as well.

Among the tenets of posthumanistic thinking is the vision of the human as embodied and embedded in cultural and technological environments with intelligence and agency distributed between humans and nonhumans (Hollinger 2009). For Marianne van den Boomen, the cyborg metaphor is characteristic of the fluidity of new media and much more sophisticated than neologisms like *prosumers* or *produsers* that also stress "the fusion of producer-consumer and producer-user [...]". The cyborg metaphor "covers the connateness of cybernetics and organic life, humans and machines, nature and culture, science and fiction" (2014, p. 153, note 131). Rosi Braidotti (2013) sees posthuman subjectivity as a tendency inherent in human and nonhuman living systems alike "to affiliate with other living systems to form new functional assemblages" (Roden 2013, n.p.) and argues that a new notion of the self as networked with capital and communication technologies across real and virtual contexts has emerged, discarding old modernist idealisation of the subject (Goodley 2014).

In the new information ecosystem, human beings can be described as interconnected informational organisms, or *inforgs* (Floridi 2014) and "surveillance is generative of bodies (of both the biological and post-biological variety) with surveillant technologies and surveillance bodies locked in a relationship of mutual interdependence" (French and Smith 2016, p. 15).

Ethical. The changing anthropological paradigm illustrated above raises a number of questions about the meaning of agency in a posthuman age. To begin with, the Council for Big Data, Ethics, and Society remarks that there are "disjunctions between big data research methods and existing research ethics paradigms" (Metcalf et al. 2016, p. 2) and warns that "epistemic conditions that were baked into research ethics regulation no longer hold in light of big data methods of knowledge production" (p. 3). Ethical challenges are articulated at different levels, "the ethics of data, the ethics of algorithms and the ethics of practices" (Floridi and Taddeo 2016, p. 3). It is largely a new scenario in which the ethical responsibilities of working with big data embrace a large spectrum of behaviours and practices that still await redefined ethics codes, for example in the areas of data generation, collection, mining and analysis (Zwitter 2014; Metcalf and Crawford 2016; O'Connell 2016). At the level of ordinary people, big data management raises a number of complex (bio)ethical issues related to privacy, ownership, confidentiality, transparency and identity. Data analytics techniques may be innocuous on their own, but can potentially lead to predictive privacy harms or data discrimination when distinct sets of highly sensitive data are combined. In the public sector, the question of Big Data and accountability is of primary concern to citizens, who should protect themselves against undue profiling, and to policy makers (Bass 2015), who should make sure that stringent privacy policies are put in place and properly enforced, also in sectors that are not necessarily familiar with data ethics. In a nutshell, "Big Data the term is now firmly entrenched, Big Data the phenomenon continues unabated, and Big Data the discipline is emerging" (Diebold 2012, p. 5), with several open issues that need to be addressed.

1.4   CONTENT DESCRIPTION AND METHODS

The standpoint of this work is the critical engagement with big data from the position of the humanities and social sciences. What has been outlined so far serves the purpose of placing the big data debate into the wider societal context and highlighting the interdisciplinary academic discussions that surround it. In what follows, the topic of big data will be primarily explored by means of a linguistic and discursive approach whose aim is to add the specific contribution of language studies to the current scholarly output. Except in the case of metaphors (Lupton 2014; Puschmann and Burgess 2014; van den Boomen 2014; Watson 2014) and, even then, mostly from the perspectives of digital sociology and media studies, this has not been extensively attempted yet.

Though big data as an innovative socio-technical construct is affecting the whole spectrum of knowledge, this investigation has prioritised three topics—big data in the global news media in English, big data and healthcare in news discourse, and the effort of European data protection legislation to come to terms with the normative issues and ethical implications of the digital ecosystem. Extrinsic and intrinsic reasons justify this kind of articulation.

First of all, this work originates from an Italian nationwide research project on knowledge dissemination, with ethics as a subtopic, in which sensitive issues in contested domains have been the focus of prolonged investigation. Second, within the domain of knowledge dissemination, the news media exercise strong influence as knowledge brokers in the field of science and technology. Journalists, bloggers, academics, freelance writers and critics work to bridge the knowledge gap between technological progress and the general audience, recontextualising and reconceptualising expert opinion from different societal sectors (Bondi et al. 2015; Watson 2016) by means of strategies of popularisation. This makes of the representation of big data in the news media a privileged point of departure to understand how technological innovation is received by the general public. Thus, a linguistic and discursive approach to news discourse can turn out to be very insightful in pinning down the ambivalent representation of data science as hype and hope but also a threat. This perspective can shed light on a societally relevant set of issues that have migrated from corporate communication and academia to the news media and institutional discourse: "Big Data discourse needs to be studied because popular technological terms and phrases play a constructive role in the way collective visions and dominant understandings about technology emerge" (Shaw 2015, p. 10).

In light of the above, Chapter 2 centres on big data in the global news media in English, investigating an ad hoc news corpus whose main topic is big data in digital innovation. First, it describes the outlines of the debate by focusing on keywords and their keyness in order to identify the most salient subtopics. It then discusses the emergence of new "data" compounds (e.g. *data-driven, data mining, big data, fake data, raw data*), the semantic prosody of collocations, and the repeated use of figurative language (like "data is the new oil"[11] or the "weaponisation of data"[12]) as a way of building discursive frames for readers by iteration. The second part of the chapter retraces interdiscursive processes in the construction of expert discourse, investigating who is legitimated as knowledgeable in the big data narrative, from which sectors of society they hail, and what views and ideologies are thereby conveyed. As for the Cambridge Analytica data breach, it observes how it managed to give momentum to the emergence of ethical reflections with regard to big data in public discourse and not just within expert communities. Before March 2018, these reflections, still in a relatively nascent state, were often obfuscated by the big data hype. Though disgracefully, the current public debate on data ethics was accelerated precisely by what the news media reported as a grossly unethical privacy breach.

The empirical scrutiny of the corpus in Chapter 2 seems to suggest that the treatment of big data in healthcare is based on different explicit and tacit assumptions. In this domain, big data analytics and machine learning are celebrated as the leap ahead in cutting-edge, patient-centric, preventive healthcare. To this end, after collecting a second smaller corpus of news items and expert commentary in popular science magazines and medical journals, Chapter 3 concentrates on the distinctive features of the representation of big data in healthcare. Keywords, concordances and the discursive encoding of agency all show the de-emphasis of individual privacy to the advantage of the overriding interests of public health and the mounting role of technology (from lifelogging devices to machine learning) in the doctor–patient interaction. Nonetheless, medical advances by means of big data leave a number of unsolved questions as regards privacy, security, collection, storage and gender and racial bias, while the legal and regulatory side of the debate is clearly a pressing issue throughout. In both Chapters 2 and 3, the chosen methodology is Corpus-Assisted Discourse Studies (CADS), which integrates Discourse and Critical Discourse Analysis with the exploratory findings obtained by the application of corpus linguistics techniques to specifically designed corpora of relevant materials (Partington et al. 2013; Baker and McEnery 2015), in order to focus on the linguistic and discursive make-up of texts.

Chapter 4 deals with a number of novel and complex ethical issues related to privacy as a basic human right and data protection, issues that are perceived as especially impellent in the wake of the recent data breaches. It focuses on the discourse analysis of the General Data Protection Regulation. The GDPR, which was approved by the European Parliament on 14 April 2016, is the new binding Regulation in the European Union, as of 25 May 2018. The relationship between the right to privacy and data protection policy is investigated in sectors that are not necessarily familiar with data ethics but need to adapt to the big data ecosystem. The inspiring principles, actors and goals of the GDPR are used to explore the discursive overlap of law and ethics and the impact generated by the EU's legal curb on data processing and repurposing. The discourse analysis of the document illustrates some of the complexities of data protection against the dematerialisation of the economy in the knowledge society and the difficulty to harmonise the multiple views of Member States on privacy. In this case, Discourse Analysis is arguably the right methodology to explore participants, processes and representations in society, starting from linguistic and discursive cues retrieved from the texts, while Critical Discourse Analysis is useful to integrate fine-grained readings of textual details into a bigger picture where ideology is questioned (Tannen et al. 2015).

Chapter 5 summarises the critical overview of the big data debate in global news, health and policy discourse and across public, corporate and academic settings. It illustrates how big data impinges on the entire spectrum of disciplines, from the hard and life sciences to the social sciences and humanities, posing unexplored ethical and methodological issues. One of the main questions is how to accommodate correlation with causation in big data analytics. It argues that the use of big data does not imply the demise of theory, the death of politics, or the disappearance of the human subject to whom data custody and stewardship responsibilities are ultimately assigned. Rather, a novel data ethics is seeing the light and with it a new way of speaking with/of technology.

To conclude, when this research was still in progress, the Cambridge Analytica scandal broke out, reviving fears about illicit surveillance on the part of governments and inviting deeper reflection on civil and human rights. It is to be hoped that global political contingencies and concerns will reinforce citizen-centric practices, like greater scrutiny of social media platforms, regulation of digital markets, environmental justice activism through big data techniques (Mah 2017) and data activism interrogating the politics of big data (Milan 2017). These are just a few of the emergent research avenues in which insights from linguistics and discourse studies can add depth to the understanding of the new social configurations of the big data age.

NOTES

1. "Artificial Intelligence" was first coined by US computer scientist John McCarthy at Stanford University in 1955 in preparation for a conference on this emerging topic. The expression "Internet of Things" was invented by Kevin Ashton in 1999, while working at Procter & Gamble as a brand manager.
2. In June 2013, twenty-nine-year-old former CIA systems analyst Edward Snowden leaked sensitive information about the US and UK surveillance programmes to the media. The *Guardian*, followed by the *Washington Post*, reported that the US National Security Agency (NSA) was operating a surveillance programme known as Prism that collected the phone records of tens of millions of Americans and tapped directly into the servers of nine internet firms, including Facebook, Google, Microsoft and Yahoo. Britain's intelligence GCHQ was also involved in the accusation.
3. The Cambridge Analytica data breach was exposed on 18 March 2018. On the "Leave" side of the Brexit referendum and financed by high-powered Republican supporters in the US, the British data firm carried out psychographic profiling on up to eighty-seven million Facebook users without their consent and used it for personalised political advertising during Donald Trump's election campaign by means of automated bots (see also Chapter 2, § 2.4.2).
4. In this work, big data is always lower case except in those citations where it is capitalised.
5. To be noted, the book *Understanding Big Data* (Zikopolous et al. 2012) has been authored by a team of IBM specialists.
6. Kitchin describes big data as possessing the following seven characteristics: "huge in *volume*, consisting of terabytes or petabytes of data; high in *velocity*, being created in or near real-time; diverse in *variety* in type, being structured and unstructured in nature, and often temporally and spatially referenced; *exhaustive* in scope, striving to capture entire populations or systems ($n=$ all); fine-grained in *resolution*, aiming to be as detailed as possible, and uniquely *indexical* in identification; *relational* in nature, containing common fields that enable the conjoining of different data sets; *flexible*, holding the traits of extensionality (can add new fields easily) and *scalable* (can expand in size rapidly)" (2014b, p. 68, original emphasis).
7. New concepts frequently spark off lists of words that attempt to capture their meaning. Besides the canonical features of *volume*, *veracity* and *variety*, big data has recently been mapped by additional "v-words" (e.g. *versatility*, *volatility*, *virtuosity*) and "p-words", in the shape of quite imaginative adjectives like *portentous*, *perverse*, *polymorphous* that the news

media echo in their turn. "While useful entry points into thinking critically about Big Data, these additional v-words and new p-words are often descriptive of a broad set of issues associated with Big Data, rather than characterising the ontological traits of the data themselves" (Kitchin and McArdle 2016, p. 2).

8. The Third Industrial Revolution was defined by economic thinker Jeremy Rifkin (2011) as the merging of internet technology and renewable energy. The *Economist* covered this change in the issue that appeared on 21 April 2011, hailing the digitisation of manufacturing. The Fourth Industrial Revolution, which is building on the Third, is characterised "by a fusion of technologies that is blurring the lines between the physical, digital, and biological spheres. [...] Impressive progress has been made in AI in recent years, driven by exponential increases in computing power and by the availability of vast amounts of data, from software used to discover new drugs to algorithms used to predict our cultural interests. Digital fabrication technologies, meanwhile, are interacting with the biological world on a daily basis" (World Economic Forum 2016, n.p.).

9. Two powerful content hubs should be mentioned to illustrate how academia is deeply engaged in the big data debate. The first one is MethodSpace, "home of the research methods community", produced by SAGE Publishing and embedding the Big Data Hub, "the community dedicated to the discussion and advancement of Big Data Analysis". The second one is the LSE Impact of Social Sciences Blog, aiming at "maximising the impact of academic work in the social sciences and other disciplines".

10. Jim Gray went missing at sea on 28 January 2007. His ideas were made into the *Fourth Paradigm* book by his Microsoft colleagues and outside scientists as a tribute to Gray's contribution to knowledge.

11. "The concept is usually credited to Clive Humby, the British mathematician who established Tesco's Clubcard loyalty program. Humby highlighted the fact that, although inherently valuable, data needs processing, just as oil needs refining before its true value can be unlocked" (Marr 2018, n.p.).

12. An essay in the First Words section of the *New York Times Magazine* (14 March 2017) recounts the birth of the word "'weaponize' in the fifties as military jargon" and, as regards its current use in the media, claims that "the fabric of civilian life is now wrapped in a linguistic fog of war" (Herrman 2017, n.p.).

REFERENCES

Anderson, Chris. 2008. "The End of Theory: The Data Deluge Makes the Scientific Method Obsolete." *Wired*, 23 June. http://www.wired.com/2008/06/pb-theory.

Andrejevic, Mark. 2014. "The Big Data Divide." *International Journal of Communication* 8: 1673–1689. http://ijoc.org/index.php/ijoc/article/view/2161.

Baker, Paul, and Tony McEnery, eds. 2015. *Corpora and Discourse Studies: Integrating Discourse and Corpora*. Basingstoke and New York: Palgrave Macmillan.

Bass, Gary D. 2015. "Big Data and Government Accountability: An Agenda for the Future." *I/S: A Journal of Law and Policy* 11, no. 1: 13–48. https://kb.osu.edu/dspace/handle/1811/75432.

Big Data Hub. https://www.methodspace.com/big-data.

Boellstorff, Tom. 2013. "Making Big Data, in Theory." *First Monday* 18, no. 10, 7 October. http://firstmonday.org/article/view/4869/3750.

Bondi, Marina, Silvia Cacchiani, and Davide Mazzi, eds. 2015. *Discourse In and Through the Media: Recontextualizing and Reconceptualizing Expert Discourse*. Newcastle upon Tyne: Cambridge Scholars Publishing.

Bowker, Geoffrey C., and Susan Leigh Star. 1999. *Sorting Things Out: Classification and Its Consequences*. Cambridge, MA: MIT Press.

boyd, danah, and Kate Crawford. 2012. "Critical Questions for Big Data." *Information, Communication & Society* 15, no. 5: 662–679. https://doi.org/10.1080/1369118X.2012.678878.

Braidotti, Rosi. 2013. *The Posthuman*. Cambridge: Polity Press.

Chandler, David. 2015. "A World Without Causation: Big Data and the Coming of Age of Posthumanism." *Millennium: Journal of International Studies* 43, no. 3: 833–851. https://doi.org/10.1177/0305829815576817.

Chen, Jonathan H., and Steven M. Asch. 2017. "Machine Learning and Prediction in Medicine—Beyond the Peak of Inflated Expectations." *The New England Journal of Medicine* 376, no. 26: 2507–2509. https://doi.org/10.1056/nejmp1702071.

Council for Big Data, Ethics, and Society. http://bdes.datasociety.net.

Coveney, Peter V., Edward R. Dougherty, and Roger R. Highfield. 2016. "Big Data Needs Big Theory Too." *Philosophical Transactions of the Royal Society A: Mathematical, Physical and Engineering Sciences* 374, no. 2080: 1–11. https://doi.org/10.1098/rsta.2016.0153.

Cowls, Josh, and Ralph Schroeder. 2015. "Causation, Correlation, and Big Data in Social Science Research." *Policy & Internet* 7, no. 4: 447–472. https://doi.org/10.1002/poi3.100.

Cox, Michael, and David Ellsworth. 1997. "Application-Controlled Demand Paging for Out-of-Core Visualization." *Proceedings of the 8th Conference on Visualization '97*, 235–244. Los Alamitos, CA: IEEE Computer Society Press. https://dl.acm.org/citation.cfm?id=267068.

Crawford, Kate, Kate Miltner, and Mary L. Gray. 2014. "Critiquing Big Data: Politics, Ethics, Epistemology." *International Journal of Communication* 8: 1663–1672. http://ijoc.org/index.php/ijoc/article/view/2167.

Diebold, Francis X. 2012. "A Personal Perspective on the Origin(s) and Development of 'Big Data': The Phenomenon, the Term, and the Discipline." Second Version, PIER Working Paper No. 13–003. https://ssrn.com/abstract=2202843.

Fisher, Eran. 2010. "Contemporary Technology Discourse and the Legitimation of Capitalism." *European Journal of Social Theory* 13, no. 2: 229–252. https://doi.org/10.1177/1368431010362289.

Floridi, Luciano. 2014. *The Fourth Revolution: How the Infosphere Is Reshaping Human Reality*. Oxford: Oxford University Press.

Floridi, Luciano, and Mariarosaria Taddeo. 2016. "What Is Data Ethics?" *Philosophical Transactions of the Royal Society A: Mathematical, Physical and Engineering Sciences* 374, no. 2083: 1–4. https://doi.org/10.1098/rsta.2016.0360.

French, Martin, and Gavin J. D. Smith. 2016. "Surveillance and Embodiment: Dispositifs of Capture." *Body & Society* 22, no. 2: 3–27. https://doi.org/10.1177/1357034x16643169.

Frické, Martin. 2015. "Big Data and Its Epistemology." *Journal of the Association for Information Science and Technology* 66, no. 4: 651–661. https://doi.org/10.1002/asi.23212.

Goodley, Dan. 2014. "The Posthuman." *Disability & Society* 29, no. 5, 844–846. https://doi.org/10.1080/09687599.2014.889416.

Graham, Mark. 2012. "Big Data and the End of Theory?" *Guardian*, 9 March. https://www.theguardian.com/news/datablog.

Halavais, Alexander. 2015. "Bigger Sociological Imaginations: Framing Big Social Data Theory and Methods." *Information, Communication & Society* 18, no. 5: 583–594. https://doi.org/10.1080/1369118X.2015.1008543.

Halford, Susan, and Mike Savage. 2017. "Speaking Sociologically with Big Data: Symphonic Social Science and the Future for Big Data Research." *Sociology* 51, no. 6: 1132–1148. https://doi.org/10.1177/0038038517698639.

Hay, Tony, Stewart Tansley, and Kristin Tolle, eds. 2009. *The Fourth Paradigm: Data-Intensive Scientific Discovery*. Redmond, WA: Microsoft Research.

Herrman, John. 2017. "If Everything Can Be 'Weaponized,' What Should We Fear?" *The New York Times Magazine*, 14 March. https://www.nytimes.com/2017/03/14/magazine/if-everything-can-be-weaponized-what-should-we-fear.html.

Hollinger, Veronica. 2009. "Posthumanism and Cyborg Theory." In *The Routledge Companion to Science Fiction*, edited by Mark Bould, Andrew M. Butler, Adam Roberts, and Sherryl Vint, 267–278. Abingdon and New York: Routledge.

Kitchin, Rob. 2014a. "Big Data, New Epistemologies and Paradigm Shifts." *Big Data & Society* 1, no. 1: 1–12. https://doi.org/10.1177/2053951714528481.

———. 2014b. *The Data Revolution: Big Data, Open Data, Data Infrastructures & Their Consequences*. London: Sage.

Kitchin, Rob, and Gavin McArdle. 2016. "What Makes Big Data, Big Data? Exploring the Ontological Characteristics of 26 Datasets." *Big Data & Society* 3, no. 1: 1–10. https://doi.org/10.1177/2053951716631130.

Lohr, Steve. 2012. "How Big Data Became So Big." *New York Times*, 11 August. http://www.nytimes.com/2012/08/12/business/how-big-data-became-so-big-unboxed.html.

———. 2013. "The Origins of 'Big Data': An Etymological Detective Story." *New York Times*, 1 February. https://bits.blogs.nytimes.com/2013/02/01/the-origins-of-big-data-an-etymological-detective-story.

Lupton, Deborah. 2014. *Digital Sociology*. Abingdon and New York: Routledge.

Lycett, Mark. 2013. "'Datafication': Making Sense of (Big) Data in a Complex World." *European Journal of Information Systems* 22, no. 4: 381–386. https://doi.org/10.1057/ejis.2013.10.

Lyon, David. 2014. "Surveillance, Snowden, and Big Data: Capacities, Consequences, Critique." *Big Data & Society* 1, no. 2: 1–13. https://doi.org/10.1177/2053951714541861.

Mah, Alice. 2017. "Environmental Justice in the Age of Big Data: Challenging Toxic Blind Spots of Voice, Speed, and Expertise." *Environmental Sociology* 3, no. 2, 122–133. https://doi.org/10.1080/23251042.2016.1220849.

Markoff, John. 2009. "A Deluge of Data Shapes a New Era in Computing." *New York Times*, 14 December. https://www.nytimes.com.

Marr, Bernard. 2018. "Here's Why Data Is Not the New Oil." *Forbes*, 5 March. https://www.forbes.com/sites/bernardmarr/2018/03/05/heres-why-data-is-not-the-new-oil.

Mazzocchi, Fulvio. 2015. "Could Big Data Be the End of Theory in Science? A Few Remarks on the Epistemology of Data-Driven Science." *EMBO Reports* 6, no. 10: 1250–1255. https://doi.org/10.15252/embr.201541001.

Metcalf, Jacob, and Kate Crawford. 2016. "Where Are Human Subjects in Big Data Research? The Emerging Ethics Divide." *Big Data & Society* 3, no. 1: 1–14. https://doi.org/10.1177/2053951716650211.

Metcalf, Jacob, Emily F. Keller, and danah boyd. 2016. "Perspectives on Big Data, Ethics, and Society." The Council for Big Data, Ethics, and Society. 7 July. http://bdes.datasociety.net/council-output/perspectives-on-big-data-ethics-and-society.

Milan, Stefania. 2017. "Data Activism as the New Frontier of Media Activism." In *Media Activism in the Digital Age: Charting an Evolving Field of Research*, edited by Goubin Yang and Viktor Pickard, 151–163. Abingdon and New York: Routledge.

Nelson, Michael L. 2009. "Data-Driven Science: A New Paradigm?" *EDUCAUSE Review* 44, no. 4: 6–7. https://er.educause.edu/articles/2009/7/datadrivenscience-a-new-paradigm.

O'Connell, Anne. 2016. "My Entire Life Is Online: Informed Consent, Big Data, and Decolonial Knowledge." *Intersectionalities: A Global Journal of Social Work Analysis, Research, Polity, and Practice* 5, no. 1: 68–93. http://journals.library.mun.ca/ojs/index.php/IJ/article/view/1523.

Paganoni, Maria Cristina. 2017. "Discursive Pitfalls of the Smart City Concept." In *Media and Politics: Discourses, Culture, and Practices*, edited by Letizia Osti, Bettina Mottura, and Giorgia Riboni, 434–451. Newcastle upon Tyne: Cambridge Scholars Publishing. http://www.cambridgescholars.com/media-and-politics.

Partington, Alan, Alison Duguid, and Charlotte Taylor. 2013. *Patterns and Meanings in Discourse: Theory and Practice in Corpus-Assisted Discourse Studies (CADS)*. Amsterdam and Philadelphia: John Benjamins.

Pentzold, Christian, and Charlotte Fischer. 2017. "Framing Big Data: The Discursive Construction of a Radio Cell Query in Germany." *Big Data & Society* 4, no. 2: 1–11. https://doi.org/10.1177/2053951717745897.

Portmess, Lisa, and Sara Tower. 2015. "Data Barns, Ambient Intelligence and Cloud Computing: The Tacit Epistemology and Linguistic Representation of Big Data." *Ethics and Information Technology* 17, no. 1, 1–9. https://doi.org/10.1007/s10676-014-9357-2.

Pots, Marieke. 2014. *How Big Data Became So Big: A Media-Archaeological Study of Transhistorical Ideas about Technology & the Rising Popularity of Big Data*. Utrecht: Utrecht University. https://dspace.library.uu.nl.

Puschmann, Cornelius, and Jean Burgess. 2014. "Big Data, Big Questions| Metaphors of Big Data." *International Journal of Communication* 8: 1690–1709. http://ijoc.org/index.php/ijoc/article/view/2169.

Reischauer, Georg. 2018. "Industry 4.0 as Policy-Driven Discourse to Institutionalize Innovation Systems in Manufacturing." *Technology Forecasting and Social Change* 132, no. C: 27–33. https://doi.org/10.1016/j.techfore.2018.02.012.

Rifkin, Jeremy. 2011. *The Third Industrial Revolution: How Lateral Power Is Transforming Energy, the Economy, and the World*. New York and Basingstoke: Palgrave Macmillan.

Roden, David. 2013. "Book Review: Braidotti's Vital Posthumanism." *Humanity+ Magazine*. 18 November. http://hplusmagazine.com/2013/11/18/bookreview-braidottis-vital-posthumanism.

Shaw, Ryan. 2015. "Big Data and Reality." *Big Data & Society* 2, no. 2: 1–4. https://doi.org/10.1177/2053951715608877.

Shin, Dong-Hee, and Min Jae Choi. 2015. "Ecological Views of Big Data: Perspectives and Issues." *Telematics and Informatics* 32, no. 2: 311–320. https://doi.org/10.1016/j.tele.2014.09.006.

Tannen, Deborah, Heidi E. Hamilton, and Deborah Schiffrin, eds. 2015. *The Handbook of Discourse Analysis*. Second Edition. Oxford: Wiley Blackwell.

van den Boomen, Marianne. 2014. *Transcoding the Digital: How Metaphors Matter in New Media*. Amsterdam: Institute of Network Cultures.

van Dijck, José. 2014. "Datafication, Dataism and Dataveillance: Big Data between Scientific Paradigm and Ideology." *Surveillance & Society* 12, no. 2: 187–208. https://doi.org/10.24908/ss.v12i2.4776.

Verbeek, Peter-Paul. 2011. *Moralizing Technology: Understanding and Designing the Morality of Things*. Chicago: University of Chicago Press.

Watson, Sara M. 2014. "Data Is the New '____': Sara M. Watson on the Industrial Metaphor of Big Data." *DIS*. http://dismagazine.com/discussion/73298/sara-m-watson-metaphors-of-big-data.

———. 2016. "Toward a Constructive Technology Criticism." Tow Center for Digital Journalism White Papers. New York: Columbia University. https://doi.org/10.7916/D86401Z7.

Williams, Sierra. 2014. "Rob Kitchin: 'Big Data Should Complement Small Data, Not Replace Them.'" LSE Impact of Social Sciences Blog, London School of Economics and Political Science, 27 June. http://blogs.lse.ac.uk/impactofsocialsciences/2014/06/27/series-philosophy-of-data-science-rob-kitchin.

Woodfield, Kandy, ed. 2017. *The Ethics of Online Research*. Advances in Research Ethics and Integrity, Volume 2. Bingley, UK: Emerald Publishing.

World Economic Forum. 2016. "The Fourth Industrial Revolution: What It Means, How to Respond." 14 January 2016. https://www.weforum.org/agenda/2016/01/the-fourth-industrial-revolution-what-it-means-and-how-to-respond/.

Zikopolous, Paul C., Chris Eaton, Dirk de Roos, Thomas Deutsch, and George Lapis. 2012. *Understanding Big Data: Analytics for Enterprise Class Hadoop and Streaming Data*. New York: McGraw-Hill.

Zwitter, Andrej. 2014. "Big Data Ethics." *Big Data & Society* 1, no. 2: 1–6. https://doi.org/10.1177/2053951714559253.

CHAPTER 2

Big Data in the News Media

Abstract This chapter discusses how 'big data' has become a catchphrase in the technology section of the news media. Through the synergic tools of Corpus-Assisted Discourse Studies (CADS), it identifies the news values and linguistic and discursive features in global big data coverage in English and elicits what kind of rhetoric is emerging. The big data narrative is rife with metaphors and novel lexical compounds. Keywords, concordance lines and collocations construct a mixed semantic prosody that takes a marked negative turn after the recent instances of data leaks and privacy violations. Finally, the analysis focuses on the strategies deployed in the construction and dissemination of expert discourse about big data by observing the processes of reconceptualisation and recontextualisation of knowledge that are activated in its argumentation.

Keywords Big data · Corpus-Assisted Discourse Studies (CADS) · Expert discourse · Knowledge dissemination · News media

2.1 The News Media and Knowledge Dissemination of Big Data

One of the roles of the news media is to serve as a bridge between experts and the lay public by enabling the circulation of ideas in order to facilitate the reception of scientific and technological innovation that pervades many aspects of society. For the majority of people, the news media

M. C. Paganoni, *Framing Big Data*,
https://doi.org/10.1007/978-3-030-16788-2_2

21

are the preferred channels through which they can stay informed about scientific and technological advances, from which they would be otherwise excluded. To address this perceived gap, researchers today are invited to communicate scientific activities and news to wide audiences in the public sphere. Besides, topics involving technologies and new devices attract readers to the media, generating revenues that news publishers need to keep afloat in the highly competitive media ecosystem (Caulfield 2004).

In other words, the process of knowledge dissemination across communication settings—from researchers in science and tech experts to the general public—and, consequently, the choice of ad hoc linguistic and discursive strategies are shaped by economic, social and cultural variables that go beyond the strictly informative and educational mission of journalism. Epistemological claims about what counts as knowledge and its uses are rife with explicit, as well as tacit, ideological implications.

Big data today is a controversial topic, still elusive for non-tech-savvy readers, and articulated around a number of contested or contestable meanings. According to the *New York Times*, it was the year 2012 when the term "big data" became mainstream as a featured topic at the World Economic Forum in Davos, therefrom making it to the headlines (Lohr 2012), where it has increasingly appeared as a catchphrase of futuristic appeal. To date, claims about big data in the news media are conflicting, and a shared view has not emerged out of a number of opinions in favour and against it. Big data, for example, is represented as central to business success because of consumer profiling, it is required in the implementation of smart city initiatives, and announced as a big promise in healthcare for its predictive potential. At the same time, especially after the recent data breach scandals, the use of big data raises several questions about privacy, consent, non-discrimination and accountability that see all the actors involved—from tech industries to ordinary citizens—faced with new opportunities and dilemmas. A large spectrum of behaviours and practices still await redefined ethics codes, for example in the areas of data generation, collection, mining and analysis (Metcalf and Crawford 2016).

In this chapter, the linguistic analysis of the big data debate moves from the endpoint in the knowledge dissemination process, the news media, with the aim to illustrate its main arguments. The traditional notion of news values (Galtung and Ruge 1965) is refreshed by a critical approach that sees them as discursive constructions whose ideological aspects are textually realised and can thus be investigated from the perspective of Critical Discourse Analysis, fruitfully integrated by corpus linguistic techniques (Bednarek and Caple 2014, 2017). More specifically, by applying a mixed-methods approach, the analysis intends, first, to describe the new kind of rhetoric that is emerging and shaping social imaginaries with the help of the main linguistic and discursive features that recur in media coverage of big data. It then reflects on how the topic is ideologically framed and verifies to what extent the epistemological, anthropological and ethical issues anticipated in the previous chapter are thematised. Finally, it focuses on the strategies deployed in the construction and dissemination of expert discourse about big data by observing the processes of reconceptualisation and recontextualisation of knowledge (Bondi et al. 2015) that are activated in its argumentation. To this purpose, specific research questions are addressed: Who is represented as the actors of the big data narrative (i.e. academics in science and the humanities, the corporate world, ordinary citizens)? What are the main views that are voiced? What kind of ideologies are thereby conveyed?

2.2   MATERIALS AND METHODOLOGY

The aim of the project was to search for the linguistic and discursive ways in which big data has been framed in the news, given their major role of global media in knowledge dissemination. To this purpose, an electronic study corpus ("Big Data corpus", or BD) was collected from online news sources from the UK, US and Australian quality press, by searching for "big data". The time span taken into consideration went from January 2016 to the last week of March 2018 when the Cambridge Analytica scandal broke out,[1] introducing new powerful elements in the narrative and altering the news frames used in reporting on the topic. The set of pragmatic decisions that presided over the corpus building phase will be illustrated below in detail.

As the name itself reveals, the Big Data corpus was manually created around a very specific technological topic, starting from preliminary web searches and researcher-driven hypotheses. It soon became evident that big data in healthcare was handled differently in the news media and that the health subtopic required a different set of premises, especially as regards the notions of privacy and public interest. This led to the compilation of a second corpus about big data in the biomedical field that will be discussed in a separate chapter. After Cambridge

Analytica, moreover, a third small corpus was built to investigate changes in news framing. Each of the three small study corpora that resulted can be described as topic-specific and user-defined (Bednarek and Caple 2014, p. 137) and contains comprehensive news coverage over an adequate time span. In other words, in the construction of specialised small corpora, criteria of representativeness are researcher-driven and based on the specific purposes of the intended analysis (Kilgarriff and Grefenstette 2003). Finally, because of their limited amount of information, small corpora lend themselves to qualitative analysis better than large corpora (Koester 2010).

Reasons of digital accessibility were prioritised, especially in light of media convergence and content sharing between digital and print. The majority of online news items (news stories, columns, editorials, op-eds, reviews, blog posts), which were all saved in individual plain text files, were retrieved from LexisNexis Academic,[2] the rest from the Europresse portal and Google News. During corpus building, news items were individually scrutinised to check whether they were pertinent subject-wise. A preliminary manual check was carried out in order to exclude texts in which the topic of innovative technology was just a contextual prop. The purpose of this was a way to avoid sporadic and scattered references to big data as a rhetorical device (e.g. "in the age of big data"), as this was felt to be tangential compared to more in-depth coverage of the theme.

The UK titles taken into consideration are the *Economist* (the only weekly publication), the *Financial Times*, the *Guardian*, the *Independent*, the *Telegraph*, the *Times* and *Sunday Times*. The US broadsheets gather the *Los Angeles Times*, the *New York Times*, *USA Today* and the *Washington Post*. The *Age*, the *Australian*, the *Australian Financial Review*, the *Canberra Times*, *Guardian Australia* and the *Sydney Morning Herald* are the Australian sources. The analysed full-text stories form an overall news corpus of 297 published items. The three subcorpora contain 95 (UK), 93 (US) and 109 (Australia) news items each, with the noticeable presence of a few duplicates in the Australian press, which makes the three corpora basically of the same size in terms of number of items. The explanation for this is that Australia is characterised by one of the highest levels of media ownership concentration in the world. Fairfax Media, which merged into Nine Entertainment in December 2018, owns the *Age*, the *Australian Financial Review*, the *Sydney Morning Herald* and the *Canberra Times*, which explains the practice of copy sharing between the group's mastheads.[3] The duplication of these news stories, which was taken as a signal of their newsworthiness, was automatically corrected by the corpus query system adopted for the analysis, i.e. Sketch Engine (Kilgarriff et al. 2014), as "substantial over-representation of subsets [...] may overwhelm meaningful topical patterns" (Schofield et al. 2017, p. 2737). At that point, the corpus amounted to 305,010 words, with a 7.65 type-token ratio (TTR) as a raw measure of lexical diversity (Baker 2006).

Some of the selected sources taken into consideration have international editions (e.g. the *New York Times*) or country-specific editions (*Guardian Australia*), which results in a mix between global coverage and country-specific perspectives. The presence of a technology section on each and every news website was also checked.[4] In other words, the corpus contains tech pieces, often in the form of editorials, op-eds and columns, that are intended for global readers of science and technology alongside a variety of news stories in which the topic is reported in association with a local newsworthy event or issue, or in connection with a given country or geopolitical context (i.e. the European Union).

Since Cambridge Analytica, growing concerns about data privacy and the misuse of information for political reasons have surfaced as additional and compelling evidence in the news media. For this reason, a smaller corpus ("Big Data after Cambridge Analytica", or ACA) amounting to 85 news items was collected from 27 March 2018 to 30 June 2018 in order to enrich the investigation of the main corpus by illustrating the discursive turning point that has taken place in big data storytelling. This second corpus contains 135,186 words, with a 6.81 type-token ratio. This figure is slightly lower than for the first corpus, presumably by reason of the greater specificity of its main topic—the Facebook-Cambridge Analytica data scandal—which results in the more frequent repetition of the same words and similar phrases (Baker 2006, p. 52).

The chosen methodology was Corpus-Assisted Discourse Studies (CADS), which combines the theory and methods of both Corpus Linguistics and Discourse Analysis. Through this mixed-methods approach, linguistic analysis is supported by a quantitative basis of observable relationships and patterns in language use while still addressing discourse structures and obtaining insights into the ideology of text. Over the years, the methodology and its strengths have gained popularity (Garzone and Santulli 2004; Partington 2004a; Baker 2006; Baker et al. 2008; Partington et al. 2013; Baker and McEnery 2015).

By using corpora and software, Corpus Linguistics allows to query large amounts of text through computer-enhanced tools, test hypotheses quantitatively and elicit broad discursive patterns through staple techniques like word frequency, keywords with their keyness score, and concordances.[5] Expanded concordance lines give access to context, while the analysis of collocates and phrases surrounding lexical items associated with the topic under analysis (Baker 2006; Baker et al. 2008; Baker and McEnery 2015) may shed light on the ideological dimension of language (Stubbs 1996, 2001). In its turn Discourse Analysis, which explores how actors, processes and representations are encoded in language, can benefit from the empirical findings of Corpus Linguistics (Thornbury 2010). As for news discourse, the approach allows to expose "how media texts might be *repeatedly* framing issues or events that are reported over a significant period of time" (O'Halloran 2010, p. 563, original emphasis), positioning descriptive and interpretive findings within a larger social context and, possibly, being able to assess their non-obvious ideological import by way of Critical Discourse Analysis.

The Corpus Linguistics software used for the analysis is the latest version of the Sketch Engine software (Kilgarriff et al. 2014). Processing text with Sketch Engine provides cues for keywords and cross-associations of collocations, thus facilitating the retrieval of complex discourse patterns whose interpretation is typical of Discourse and Critical Discourse Analysis.

2.3   QUERYING THE BIG DATA CORPUS

The first step of the analysis consisted in a number of probes into the Big Data corpus to identify high frequency words and keywords in order to assess textual material "from the point of view of topical cohesion" (Thornbury 2010, p. 273). Since there are several current and potential applications for big data, the aim was to make a preliminary recognition of the knowledge areas and the societal actors foregrounded in news coverage.

Thanks to the Word List tool, a frequency lemma list was generated and arranged in descending order.[6] Thirty content lemmas were isolated out of the top one hundred entries, manually checking the occurrence of possible word forms for each lemma in the corpus (Table 2.1). Although the cut-off point (at thirty in this case) and the choice of vocabulary were discretionary, the decision was justified by the aim to

Table 2.1 Descending lemma frequency list in the Big Data corpus

No.	Lemmas in the BD corpus	Frequency
1	data	2784
2	use/uses/using/used	1146
3	technology/technologies/tech	978
4	company/companies	965
5	big/bigger/biggest	934
6	people	915
7	new/newer/newest	879
8	information	785
9	business/businesses	570
10	system/systems	547
11	customer/customers	522
12	car/cars	467
13	digital	447
14	service/services/serviced	439
15	government/governments	434
16	change/changes/changing/changed	405
17	Facebook	395
18	bank/banks/banking	367
19	internet	346
20	consumer/consumers	343
21	firm/firms	342
22	human/humans	332
23	market/markets/marketing/marketed	317
24	Google	309
25	privacy	299
26	industry/industries	288
27	algorithm/algorithms	275
28	computer/computers	271
29	risk/risks/risking	239
30	health	237

gather empirical quantitative evidence to back up the following qualitative analysis. Arguably, this list of adjectives and nouns (which includes verb forms in the case of verbal nouns) sums up the basic vocabulary of big data coverage and highlights recurrent subtopics.

With *data* on top as expected and *privacy* comparatively much lower in rank, the selection of high-frequency content words can be read as a snapshot of the technologies, actors and crucial issues in big data analytics, in particular that of data protection as a right at risk. In dealing with the repercussions of digital innovation, the media narrative is polarised between the promised land of new technology (*the use of big data*) and the quicksand of *privacy risks.*[7] In between are a number of terms related to the technological domain and the information superhighway, as well as references to Silicon Valley tech giants, *banks, companies, firms* and *industries,* and collective actors like *people* and *governments, customers* and *consumers* occupying the middle ground.

Undoubtedly, the big data topic is perceived first and foremost as technological innovation (the hits of the lemma *technology* are added to the occurrences of *tech*) and societal *change,* deeply affecting the business and financial sectors (banks and fintech businesses, companies, industries), auto manufacturers (with futuristic driverless cars) and the service sectors, i.e. the entire economic *system.* The high frequency of *use* (both as a verb and a noun) is a significant clue that alludes to the human intentionalities on which big data analytics ultimately relies (Portmess and Tower 2015), despite more radical claims that see machine learning and artificial intelligence as being independent of human reason. *Information* and the *internet* are the pillars of a reality which is increasingly online and *digital* (Gurrin et al. 2014; Ball et al. 2016; Lanzing 2016) in which, anthropologically, *humans* have become "inforgs", or informational organisms (Floridi 2011).

Moreover, the results highlight the digital duopoly of *Facebook* (395 occurrences) and *Google* (309), against the 182 occurrences of the e-commerce giant *Amazon.* In actual fact, the topicality of the three platforms seems to correspond to their respective positioning in the digital market. According to the most updated report released by the Reuters Institute for the Study of Journalism, Facebook and Google are the Big Two of internet advertising, while Amazon represents their greatest threat. With its shopping search engine and live streaming of NFL games, it is estimated to be dominating the hardware market everywhere except for China (Newman 2018).

Algorithms are central to the commercial success of platforms as they can recognise patterns in data, predict consumer behaviour on the basis of previous purchases and thus provide customised suggestions. In several key societal domains like finance, algorithms power machine learning and make decisions instead of people, while predictive algorithmic analytics will become a routine part of clinical practice in the future. Finally, *health* is increasingly affected by the use of big data. Practices of self-tracking and lifelogging are made available to ordinary people by means of smartphone apps and digital devices and encouraged by movements like the Quantified Self (Gurrin et al. 2014; Ajana 2017).[8]

The following step of the analysis saw the extraction of keywords and bigrams by keyness, a score computed by a statistical formula that shows what words are typical of the analysed corpus because of their higher frequency against a reference corpus (Scott 1997), here the enTenTen13 corpus of general English.[9] Described as "simple maths for keywords" by the founder of Sketch Engine, the statistical computation formula includes "a variable which allows the user to focus on higher or lower frequency words" (Kilgarriff 2009, p. 1). Here this N variable, also known as the "simplemaths parameter" (Kilgarriff 2012, p. 5), was set for mid-frequency words ($N=500$, i.e. from rank 500 upwards), thus excluding rare words.[10] The procedure of extracting key keywords and bigrams was performed to guide the analysis towards the most prominent concepts in the study corpus, with the aim of retrieving a more detailed description of its "aboutness" (Scott 1999; Baker 2004; Scott 2010) that had been only partially outlined in the frequency word query.

Twenty-four content lemmas and bigrams (nominal lexical collocations in this case), ordered by keyness, out of the top fifty, again with an arbitrary cut-off, are listed in Table 2.2. Bigrams were very useful to better circumscribe the semantics of a few keywords, whose social and cultural implications could not just be captured by automatic extraction, showing meaning relationships between them and thus preparing the ground for the analysis of discourse topics. For example, the preferred meaning of *information* appears to be *personal information*, whose salience is then compounded by the near synonyms *private/sensitive information* and *personal/sensitive data* and whose social meaning is credibly related to the domain of privacy regulation and law (again, a separate chapter in this book).[11] Names such as *Trump, Cambridge Analytica* and *Equifax* and the bigram *presidential election* were not considered because, at this point, it was decided to place the focus on general topics and not on single instances of data breaches that would be discussed separately in order to avoid a lopsided analysis, leaning too much towards the dystopian features of the big data narrative. This discretionary decision should be seen as part of the fine balance in keyword analysis between "the computer's very blindness" (Scott 2010, p. 45) and human interpretation.

Table 2.2   Keywords and bigrams (by keyness) in the Big Data corpus

No.	Keywords	Keyness	Bigrams	Keyness
1	data	10.57	artificial intelligence	1.69
2	technology	4.22	chief executive	1.40
3	big	3.30	personal information	1.33
4	digital	2.96	virtual reality	1.21
5	Facebook	2.75	cyber security	1.18
6	company	2.74	industrial revolution	1.17
7	information	2.63	big data	1.16
8	privacy	2.50	credit card	1.16
9	customer	2.49	fake news	1.15
10	algorithm	2.48	new technology	1.14
11	firm	2.36	machine learning	1.13
12	Google	2.24	digital age	1.12
13	car	2.21	smart home	1.12
14	bank	2.20	early intervention	1.11
15	internet	2.09	cloud computing	1.10
16	government	2.07	identity theft	1.10
17	intelligence	1.99	digital technology	1.09
18	Amazon	1.92	augmented reality	1.08
19	human	1.92	data collection	1.08
20	social	1.88	digital disruption	1.08
21	computer	1.88	driverless car	1.08
22	system	1.87	smart city	1.08
23	security	1.84	credit reporting	1.08
24	breach	1.78	data breach	1.08

At this exploratory stage, the findings show the remarkable impact of big data in several sectors of society, but confirm the discursive polarisation of the topic. Digital innovation holds tremendous promise for mankind and, in some cases, is already part of the everyday experience of consumers. Take *artificial intelligence*, for example, with AI voice assistants based on artificial neural networks like Alexa, Siri, Google Assistant, Viv, and Microsoft's Cortana on smartphones and personal computers and in cars. However, it can be disrupted by *fake news*, *data breaches*, and *identity theft*, among other things.

If dissemination is inadequate and the knowledge gap between experts and the general public is too wide, this also increases value polarisation, which surfaces in the mixed semantic prosody of specific lexical units (Partington 2004b). Two cases in point are represented by the semantic prosody of *algorithm* and *driverless car*, both key subtopics in the study corpus. Quite understandably, algorithmic decision-making—a set of automated procedures applied in machine learning from big data—and algorithmic bias tend to be largely impenetrable topics to the general public. As for self-driving vehicles, though quite a popular notion, they seem to still be decades away from actual implementation.

In order to articulate this point, empirical evidence was gleaned from concordances generated by Sketch Engine. In Corpus Linguistics, concordance lines are useful to visualise how keywords and multi-words are inserted in a text and shed light on semantic associations (Stubbs 2001). "On the horizontal line, concordances display instances of language use, whereas if scanned across the vertical line, they can reveal repeated co-occurrences (or collocations) in the language system" (Koteyko 2010, p. 657). In the analysis of news discourse in particular, the concordance function is useful to reveal patterns that convey "a sense of what regular readers would be exposed to" (O' Halloran 2010, p. 569).

Selecting the wildcard *algorithm** as the node word by reason of its keyness, concordance lines in the KWIC format[12] were manually inspected to examine it in context, both horizontally and vertically. It was then possible to extrapolate the following examples from the extended co-text (Table 2.3).

These examples were sampled on purpose to illustrate how the knowledge potential of the algorithm, in itself a mathematical construct, is often overridden by a range of negative emotions. Since algorithms self-learn from correlating data, their predictive models may embed bias, lead to self-fulfilling prophecies that reinforce the flaw in their design, i.e. racial, gender, age and disability stereotypes, and thus affect fundamental rights. Central to the commercial success of online platforms, algorithms can recognise patterns in data, predict consumer behaviour on the basis of their previous purchases, and thus provide customised suggestions. In several key societal domains like finance, algorithms power machine learning and make decisions instead of people. All this conjures up "dystopian fears of algorithmic control" (Ames 2018, p. 3) and prompts some ethical questions (Mittelstadt et al. 2016). To some extent, this holds true for *driverless car** as well (Table 2.4). Despite the opportunities they may provide to disabled people, self-driving vehicles raise questions about human agency and ethical and legal concerns.

Such Examples illustrate how "[t]he features of the scientific and technological paradigm that big data claims to stake out are still in a period of interpretive flexibility and of ongoing contestation over

Table 2.3   Edited concordance lines of *algorithm**

There are some decisions that an	*algorithm*	cannot make
It's tempting to think that if we just run the	*algorithm*	, slice the data and dive into the analytics, we'll gain the insights and intelligence
But when an	*algorithm*	is used as the basis for a decision it is done in a semi-automatic way
In other words, the Facebook	*algorithm*	picked a side—it's not neutral.
bias or pattern that becomes obvious when the	*algorithm*	is applied to it.
Using a secret	*algorithm*	, Sesame credit constantly scores people
the new generation of AI	*algorithms*	can perform a wide variety of involved tasks
This shift, in which automation and	*algorithms*	move into every industry
AI software creates biometric	*algorithms*	of facial features.
Without a doubt,	*algorithms*	are faster than our conscious thinking
He cautions against blind faith in	*algorithms*	or in data
often uncritical in their adoption of	*algorithms*	and analytics to measure performance
Analysts often refer to	*algorithms*	as a black box
To find out whether pricing	*algorithms*	manipulate markets or even collude
The tyranny of	*algorithms*	is part of our lives
and endlessly panic about how the	*algorithms*	will rate us tomorrow
He said the deep learning	*algorithms*	which drive AI software are "not transparent"
a frightening look at how	*algorithms*	are increasingly regulating people
The inherent bias in AI	*algorithms*	by training a machine
an ongoing debate about the biases of the	*algorithms*	used. Will they challenge us enough?
biases, research reveals machine learning	*algorithms*	are picking up deeply ingrained race and gender prejudices

their exact meanings and values" (Puschmann and Burgess 2014, p. 1691).

In order to shed further light on the discursive framing of big data, the Word Sketch tool was exploited to retrieve all those clusters (bigrams and trigrams) in which *data* is head or premodifier. Top automatic hits extracted by reason of their frequency were integrated by additional "data" noun phrases and compounds that were manually retrieved from the corpus, thus guiding the critical approach towards the

Table 2.4 Edited concordance lines of *driverless car* *

insurance companies have to decide, again, whether the human inside the	*driverless car*	could have prevented an accident.
In terms of the wow effect, however, it's hard to go past the Google's	*driverless car*	Make no mistake: this is coming, and fast.
Google's	*driverless car*	isn't designed to let humans take control.
a fatal accident and there is an occupant in the	*driverless car*	? What happens if there is a lawsuit?
physically incapable of doing those things without the help of a	*driverless car*	How should insurance companies, car makers and the law treat those folks?
like machine learning, drones and	*driverless cars*	, will have similar effects—but on many more people
zero road deaths by 2020 and Volvo sees	*driverless cars*	as the solution.
advances in AI to develop his Tesla	*driverless cars*	. However, he believes the technology represents an existential threat to mankind
change how we think of human drivers, too.	*Driverless cars*	could allow those who have difficulty driving—such as the elderly or the blind
To be sure,	*driverless cars*	may be safer, potentially saving millions of people's lives every year.
Hackers and terrorists may turn	*driverless cars*	into weapons.
crashes where its cars are at fault.	*Driverless cars*	could therefore expand the range of Google's future legal responsibilities
letting humans take over actually makes	*driverless cars*	less safe, because a passenger
with even bigger ethical quandaries. Take	*driverless cars*	, which I believe will have a huge, unknowable impact on society.
disabled people won't be able to take advantage of	*driverless cars*	unless someone with a valid license is with them

qualitative dimension. Results appear in the table, listed in alphabetical order (Table 2.5).

Some clusters in both columns are premodified, or headed, by a term belonging to the lexis of business and finance, as further proof of the corporate origin of big data (*data assets, banking data, data broker, compliance data, credit data, data revenues*). Several are "metaphorical in nature" and useful as "contextualizing or framing devices"

Table 2.5 "Data" noun phrases and compounds in the Big Data corpus

No.	Data as head	Data as premodifier
1	Account data	Data analysis, analytics
2	Anonymised big data	Data assets
3	Banking data	Data attacks
3	Big data	Data availability
4	Biometric data	Data-based, -driven, -focused, -powered
5	Borrower data	Data breach/leak
6	Client data	Data brokers/brokerages
7	Compliance data	Big data capabilities
8	Consumer transaction data	Data capture/retrieval/storage
9	Corporate data	Data centre(s)
10	Credit data	Data cloud
11	Crime data	Data collection
12	Customer data	Data company
13	Consumer data	Big data consultancy
14	Digital data	Data consumption trends
15	Encrypted data	Data creation
16	Environmental data	Data crunchers
17	Equifax data	Data czar
18	Facebook data	Data deluge/stream/well
19	Fake data	Data deregulation
20	Financial data	Data digitisation
21	Fitness data	Data economy
22	Government data	Data engineers
23	Genetic data	Data ethics
24	Health/healthcare data	Data evangelist
25	Large scale data	Data exhaust
26	Lifestyle data	Data experts
27	Location data	Data flows
28	Machine data	Data giants
30	Medical data	Data governance
30	Medicare data	Data hoard
31	Open data	Data-hungry/-poor/-rich
31	Patient data	Data hygiene
32	Performance data	Data insights
33	Personal data	Data laundering
34	Political data	Data legislation
35	Private data	Data linkage
36	Private sector data	Data management
37	Public sector data	Data market
38	Raw data	Data matching
39	Real-time data/rich data	Data mining/miners
40	Search data	Data points

(continued)

Table 2.5 (continued)

No.	Data as head	Data as premodifier
41	Sensitive data	Data policy
42	Sensor data	Data portability
43	Small data	Data processes
44	Smart data	Data protection/data protection rules
45	Traffic data	Data quality
46	Transactional data	Data records
47	Unstructured data	Data refiners/refineries
48	User data	Data revenues
49		Data revolution
50		Data science/scientists
51		Data sharing
52		Data sources
53		Big data strategy
54		Data streams
55		Big data technology
56		Data transfers
57		Data trove
58		Data use
59		Data volume
60		Data vulnerabilities

(Koteyko 2010, p. 656). Among them, a few collocations of novel coinage like *data crunchers* and *data laundering* are noticeable, a phenomenon which illustrates how technical and scientific communication can creatively expand its own discourse to make room for new concepts and meanings by way of neologisms. *Raw data* and *data refineries* frame big data as crude oil that needs to be transformed into a useable product. *Smart* and *sensitive data* is an organic metaphor which turns its object into a living entity with its *hygiene* and *vulnerabilities*, somewhat conveying the notion that big data is an intelligent and sentient being.

Data deluge, stream, trove and *well* describe the unprecedented volume, mobility and richness of digital information. These same features are stressed by recurrent collocational patterns: *the enormous/fresh/huge/impressive/massive/vast/amount(s) of data, the deluge of data, the explosion of big data, the flood of data, the flow(s) of data, masses of data, mountains of data, oceans of data, the outpouring of data, piles of data, the power of data, the puffs of data, the sheer quantity of data, (huge) quantities of data, reams of data, slivers of data, streams of data, vast treasuries of data, the cascading volumes of personal data, the sheer volume of data, the swamp of big data, data's massive yields.*

Figurative expressions fall into two main conceptual metaphors (Puschmann and Burgess 2014, p. 1691) that will be illustrated below. The first one sees in big data a force of nature to be controlled (examples 1 and 2).

(1) In the nearly 10 hours that it takes a Boeing 737 to fly from Sao Paulo to New York, its twin engines will transmit <u>a flood of digital data</u> roughly equivalent to 15,000 Blu-ray movies. <u>That electronic Niagara</u> provides a continuous readout on the jet's performance, giving ground-based technicians a head start on unanticipated repairs and reducing costly down time. (*Washington Post*, 28 November 2017)

(2) *(Michael Dell, Dell Technologies)* "The internet of things will become the internet of everything. We will soon have 100bn connected devices and then a trillion, and <u>we will be awash in rich data</u>. More importantly, we will be able <u>to harness</u> that data with unprecedented computing power to drive radical progress through our global society". (*Sunday Times*, 15 October 2017)

In this case, the "extravagant materiality of big data" is conveyed by "vivid metaphors of fire, water and cosmos" while "data analytics […] is cast as survival knowledge that conquers floods, orders chaos, and stems the information apocalypse" (Portmess and Tower 2015, p. 3), once it has been unleashed.

The second conceptual metaphor equates big data with nourishment/fuel to be consumed. Consequently, information ("vast swathes of information") and "the riches of data" are a commodity to exploit and one of the most valuable assets in the world. Data is the new oil, the black, digital gold of today, the gold dust, the new air in the world of business, the lifeblood for retailers, as in examples 3–7.

(3) Data is the <u>fuel</u> of the future as well as the <u>rich soil</u> in which everything will grow. (*Sydney Morning Herald*, 20 July 2017)

(4) "Big data" is the <u>black gold</u> of today. (*Telegraph*, 14 September 2017)

(5) "We wouldn't have the current era of A.I. without the big data revolution," Dr. Li said. "It's the digital gold". (*New York Times*, 8 January 2017)

(6) Data is the "new air" in the world of business. (*Australian Financial Review*, 25 July 2017)

(7) Data is nothing less than the new lifeblood of capitalism. […] The global circulation of data, then, is really about the global circulation of capital. (*Guardian*, 1 February 2018)

However, framing big data as the new oil of the Fourth Industrial Revolution is ultimately a logical fallacy because, whereas oil is burnt, big data can be repurposed. As will be seen in the discussion of the new European regulation protecting personal data, the GDPR, the concept of purpose in big data analytics can give rise to contradictions. In the current digital gold rush, in particular, it is the question of data ownership that seems to be neglected. This same consideration applies to the use (and misuse) of big data, a range of actions that are lexicalised by verbs that include *access, aggregate, analyse, archive, capture, collect, crunch, delete, drill, encrypt, exchange, extract, filter, gather, generate, hack, harness, harvest, index, leverage, match, mine, pool, process, protect, record, refine, report, share, siphon, steal, tap, trawl, unscramble, vacuum up* and *wipe*.

Moving from extended lexical units to discourse patterns that affect the global meaning of text and news framings, we observe, once more, that the big data narrative alternates between enthusiasm and fear, and that the power of customisation in business and prediction (i.e. in crime prevention) can be flawed by vested interests and discriminating bias (examples 8–10).

(8) An artificial intelligence tool that has revolutionised the ability of computers to interpret everyday language has been shown to exhibit striking gender and racial biases. The findings raise the spectre of existing social inequalities and prejudices being reinforced in new and unpredictable ways as an increasing number of decisions affecting our everyday lives are ceded to automatons. (*Guardian*, 13 April 2017)

(9) From harnessing Big Data to track and understand customer behaviour to leveraging its cloud computing unit, Amazon Web Services, to power its global marketplace, Amazon's technology aims to squeeze value out of every bit of available data to make life easier for buyers and—sellers alike. (*Australian*, 22 April 2017)

(10) In tech, your personal data is a ripe resource for businesses to harvest in their own interests. (*New York Times*, 4 May 2017)

We also encounter "big data capabilities", "the opportunities in big data", "the predictive power of big data" that are contrasted with "fear factors about big data", "the weaponisation of big data" and "the opaque, tentacled and ballooning world of big data" and show how big data is frequently surrounded with a mixed semantic prosody.

As stated in the aims of this chapter, corpus interrogation is carried out not only to elicit the main linguistic features in the news media's big data narrative but also to investigate how the news values that are associated with it are discursively constructed. The leading assumption is that news values, which define the topics and criteria for journalists to select events, stories and ways of reporting them, exist in and are constructed through discourse and as such become a quality of text (Bednarek and Caple 2014, p. 137).[13] The newsworthiness of big data is mostly conveyed through the news values of Superlativeness, Impact and Novelty and the corresponding and, at times, overlapping linguistic devices that will be briefly illustrated below.

Superlativeness is built by means of quantifiers such as the enormous numbers involved in the age of big data (*Gigabytes, Terabytes, Petabytes, Exabytes, Zettabytes, Yottabytes…*), metaphors of intensification (*data deluge, data flood, the explosion in artificial intelligence, the tsunami of digital information*), intensified lexis ("The *sheer volume* of data being transmitted and stored is also set to *explode*"; "*endless* rows of servers in *colossal* data centers"), comparatives and superlatives ("Banks have *far more* information at their fingerprints"; "The banks are *less than* thrilled about calls for them to share the *most lucrative* data they have"). It includes verbs endowed with an intensifying meaning such as those with a prefixal use of "out" like *outpace, outperform, outrun, outstrip* and *outweigh*. Novelty is constructed by lexical occurrences like *new* as a left collocate[14] of *benchmarks, economy, technology, opportunities,* semantically related words like *change* (as noun and verb), *future* (as adjective and noun), *futuristic, innovative* and *innovation,* and comparisons with the past, as in examples 11 and 12.

(11) While business has always had to deal with change it has never had to deal with the incredible pace of change and the degree of disruption we see today. (*Age*, 2 January 2016)

(12) There has never been a time of greater promise or potential peril. (*Independent*, 28 November 2016)

Impact is conveyed by *big* in the first place, but also by lexical items like *crucial, revolution, revolutionary* and *transformation*. The use of the future tense for prediction is also functional to this rhetorical line, along with the description of "implied or explicit cause-effect relations" (Bednarek and Caple 2014, p. 156), as in examples 13 and 14.

(13) Data <u>will allow</u> businesses to better <u>understand</u> the needs of consumers, <u>improve</u> product offerings, <u>manage</u> risk and <u>reduce</u> costs. (*Australian Financial Review*, 23 March 2016)

(14) Britain is on the brink of a robotics revolution. Advances in technology <u>are unleashing</u> a new age where <u>computers handle many tasks</u> previously carried out by humans. (*Guardian*, 4 April 2016)

In sum, cultural and political anxieties about the actual use of big data repeatedly surface in news discourse, for example, by means of new compounds and metaphors that try to make sense of the phenomenon and translate it for the lay public. However, an overall very low frequency of *ethic** lemmas, synonyms and antonyms (e.g. *moral* and *immoral*) is observable in the corpus.[15] In sum, these results would seem to suggest that, before Cambridge Analytica, the risk of data breaches and leaks was not consistently framed as a profound violation of basic rights, and that human responsibility in the use of data was described as generic rather than binding. As we will see, the turning point is marked by Cambridge Analytica.

2.3.1 *After the Cambridge Analytica Data Breach*

Disrupting the relatively unchallenged storytelling about big data as most progressive and desirable in a variety of societal domains, the exposure of privacy violation raised public perception of the contradictions in the use of big data. It generated a peak of awareness about ethical breaches and loss of control over digital information, especially as concerns voter profiling and political manipulation in the "nudgital society" (Puaschunder 2017).[16] People were targeted not only on the basis of their past behaviours and preferences but also "their underlying psychological profiles" in the attempt "to interpret basic human drives and match issue messaging with personality traits" (Graves and Matz 2018, n.p.).

Since the Facebook-Cambridge Analytica revelations, the ways in which the news media have reported on big data show a decided inversion. Quite significantly, when compared with the Big Data corpus thanks to the Sketch Engine tool for comparing corpora, this second smaller corpus registers a surge in the normalised frequency (or frequency per million)[17] of negatively connoted terms. The same happens for positively connoted ones (*consent, ethics, privacy* and *transparency*) that express rights and values that were put at risk, or index social actors and legal actions at work to counteract violations (Table 2.6).

Results indicate that privacy is now a top concern, coupled with the awareness of the perceived degeneration of the big data ecosystem. Lastly, it is interesting to note that references to the General Data Protection Regulation (GDPR) increase, since the EU regulatory framework is regarded as a major attempt at making order in a novel and controversial matter.

Table 2.6 Normalised frequencies of selected lemmas in the two corpora

Lemmas	Freq/mill in the ACA corpus	Freq/mill in the BD corpus
privacy	2171.1	741.3
leak	960.7	56.2
breach	877.4	412.8
scandal	775.0	47.7
consent	691.7	56.2
legal	576.4	168.5
GDPR	486.7	36.5
regulator	461.1	168.5
investigation	422.7	115.1
transparency	377.9	89.9
ethics	339.4	25.3
misuse	333.0	25.3
lawmaker	288.2	30.9
fake (news/accounts)	275.4	126.4
transparent	230.6	30.9
ethical	211.4	73.0
complaint	211.4	47.7
disinformation	204.9	11.2
crisis	204.9	44.9
manipulation	121.7	36.5

2.4   DISCURSIVE INSIGHTS

After the retrieval of keywords and key bigrams as "pointers" to the salient meanings of text (Scott 2010, p. 56), the concomitant focus on a few selected concordances, and the discussion of "data" collocations and metaphors as framing devices, the aim of this second part is to exploit the discourse topics identified above to deepen the argumentative and ideological construal of big data coverage in the global news media (Perelman and Olbrechts-Tyteca 1969; van Dijk 1988). Big data storytelling, however, is quite heterogeneous, since expert discourse, which is largely informative and educational, alternates with breaking news coverage and social and political commentary in which technology is implicated. This thematic diversity is not amenable to the researcher-driven decision presiding over corpus design but rather is intrinsic to contemporary journalism and its information overload, multiplying intertextual and interdiscursive possibilities by way of hyperlinking. The hypertextual nature of today's news, in other words, constantly breaks up conventional journalistic categorisations and the salience criteria through which news stories have been traditionally organised in the printed media, by suggesting, instead, a labyrinth of reading paths that change with constant updates and introduce publics to a much more complex experience.

Nonetheless, for the purposes of this analysis, the main storylines that have emerged in the time span covered by the study corpus can be reasonably summarised as:

- big data in the business and finance world;
- big data in politics;
- big data in healthcare.

The main arguments in favour and against in the three cases above will be illustrated as emblematic of the polarised reception of big data. In all three scenarios, moreover, there have been major unforeseen negative events that have compromised the reputation of big data more or less severely, namely and in chronological order, the Australian Medicare data security breach in July 2017,[18] the Equifax credit data breach in September 2017,[19] and the Cambridge Analytica personal data leak in March 2018, while almost every day brings news of yet another violation, like the one that just affected PageUp, an online recruitment services organisation, adopted by major Australian companies, earlier in the month of June 2018.[20]

2.4.1   Big Data in Business

The main favourable argument for the use of big data in business is the predictive power of big data analytics, which is seen as central to business success. Big data allows companies to have far more intelligence at their disposal to make accurate decisions and, especially, predictions on their business operations. However, in an ever-increasing competitive corporate environment, businesses are unwilling to miss the "data boat", have been affected by "the big data craze", and have become intoxicated with "the drug of free data". Behind them looms digital banking that puts credit institutions in control of financial information about very large numbers of customers. In any case, the fact is that the rules of the game have changed and new actors are emerging.

> (15) Data is to this century what oil was to the last one: a driver of growth and change. Flows of data have created new infrastructure, new businesses, new monopolies, new politics and—crucially—new economics. Digital information is unlike any previous resource; <u>it is extracted, refined, valued, bought and sold in different ways</u>. It changes the rules for markets and it demands new approaches from regulators. <u>Many a battle will be fought</u> over who should own, and benefit from, data. (*Australian*, 6 May 2016)

In example 15, the management of big data technology and analytics is represented as warfare, while the use of passive structures, which elide agency, would seem to imply that control over the new resource is still undecided. Two years later (example 16), well into big data, the "data is oil" stock metaphor becomes the antecedent of a conditional clause with strong argumentative power that affectively urges readers to identify with a collective "we", sharing not only the loss of privacy but the risk of dispossession as well (example 17).

> (16) As well as our privacy and the risk to our political processes, the issue of who has the rights to our data could decide our economic future [...]. <u>If, as has been said, data is the new oil</u>, then the oil wells are in the hands of a few billionaires, and we're being pumped through the pipes. (*Guardian*, 23 March 2018)

> (17) Big data is extractive. It involves extracting data from various "mines"—Facebook, say, or a connected piece of industrial equipment.

This raw material must then be "refined" into potentially valuable knowledge by combining it with other data and analyzing it. Silicon Valley siphons our data like oil. But the deepest drilling has just begun [...] Society, not industry, should decide how and where resources are extracted. Big data is no different. The data is ours! (*Guardian*, 14 March 2018)

In the recent past, the *Economist* drew a parallel between the "robber barons" of American capitalism and today's "Silicon Sultans" (30 December 2014). More recently, "Facebook, Apple, Microsoft, Google and Amazon have together been dubbed 'the Frightful Five'" (*Canberra Times*, 19 August 2017). The "weaponisation of big data" is a recurrent topic used to denounce the instrumental manipulation of information from business to politics, not to mention the possible military uses of artificial intelligence.

At the same time, leveraging the power of big data in e-commerce allows to contrast the counterfeiting and purchasing of fake goods. Aggressive but uncritical competition, however, puts privacy at risk on an unprecedented scale in credit data where it is important to track and pursue consumers. When distinct sets of highly sensitive data are combined (healthcare, sexual orientation, religious and political views, education, employment, credit, housing, insurance, consumer, travel and exercising behaviour, retail history, welfare), deeply personal insights may be obtained, potentially leading to predictive privacy harms. The raised issue is one of data ownership, which may lead to identity theft[21] or identity fraud (as in the Equifax scandal), and also of the potential unfairness of predictive policing and of racial and gender bias in algorithmic decision making.

2.4.2 Big Data in Politics

The downsides of technology involve risks of "greater misinformation and manipulation" (Newman 2018, p. 46). In the study corpus, big data in politics means political marketing, "whispering into the ear of each and every voter" (*Financial Times*, 19 March 2018), which is part of information warfare like fake news, and whose effects are disruptive (example 18). The snapshot that best captures the situation is Donald Trump's taking office in 2016 and the rising rumours of voter profiling and "nudging" (Puaschunder 2017) that culminated in the recent Cambridge Analytica data breach that was exposed in March 2018.

Understandably, in the face of such a violation of privacy at a collective level, very critical views have been raised to expose the risk of manipulation in the absence of appropriate regulations.

> (18) Data mining and targeted messaging has become central to modern political campaigning, but the exposure of Cambridge Analytica's methodologies has raised serious doubts around the ethics and privacy law implications of the practices, with some calling it "information warfare". But there are virtually no legal limits to the use of personal data for political purposes in Australia. Australian politicians, political parties and their contractors and subcontractors have been exempt from the operation of the Privacy Act since 2000. (*Guardian Australia*, 21 March 2018)

In example 18, the use of the present perfect represents data crunching and microtargeting as an established practice in contemporary political communication that predates Donald Trump's rise to President of the United States. We know, for example, that in both the 2008 and 2012 election campaigns, Barack Obama's team managed to combine information from massive databases so as "to individually profile voters, assess if they were likely to vote and how, and how they might react to different policies and stories" (Kitchin 2014, p. 76). Predictive profiling, however, has got out of control and requires corrective action.

The mounting reaction, in the shape of "snowballing public criticisms of social media" (*Guardian*, 11 February 2018), intends to denounce the betrayal of the promises that are intrinsic to the evolution of the internet as a global technology and communication medium that should empower, and not disempower, people.[22] It should be acknowledged, however, that at least one year before the "Facebook Datagate", the media were already denouncing the erosion of democracy brought about by political marketing, and more generally hinting at "the fragility of nation states in a time of technological change" (*Financial Times*, 28 January 2018). To be noted, the *Sydney Morning Herald* and the *Canberra Times* had published critical reports on Alexander Nix, Cambridge Analytica's CEO during his trip to Australia to meet potential business partners and clients, a fact that sounds involuntarily ironic and clairvoyant in the light of the firm's attacks on transparent democratic representation. One wonders whether the "dawn of the techlash" (*Guardian*, 11 February 2018), which has developed into the full-blown ethics crisis (the "Big Tech backlash", *Financial Times*, 5 April 2018) that platforms and politics have been experiencing since the Cambridge Analytica leak will make codes of ethics more stringent.

2.4.3   *Big Data in Healthcare*

The health sector is among the societal domains that will be most affected by this new technology. Though the discursive framing of the issues surrounding big data in healthcare will be discussed in depth in the following chapter, a brief outline is anticipated here.

In a similar vein as the arguments in favour or against it that have been illustrated in relation to the world of business and politics, big data in healthcare is discursively polarised. On the one hand, it is credited with the potential to yield unprecedented diagnostic and prognostic insights into patient treatment thanks to artificial intelligence and machine learning. In addition, predictive analytics models are of fundamental assistance in health care allocation, decision-making, and delivery. At the same time, when priority is given to technocracy, the risk is to underestimate the importance of bioethical issues that are only cursorily introduced in the discussion.

On the other hand, it is undeniable that health data, which is highly personal, sensitive and potentially discriminating, is underprotected, as in the case of Medicare in Australia, which was tagged as "Mediscare" by national papers in July 2017.

(19) Despite government assurances, the hacking of Medicare cards has sparked fears over identity theft—and the safety of our health records. The Turnbull government insists there has been no major cyber security breach of its health IT systems and says "traditional" criminals—rather than sophisticated hackers—are likely to blame for a website apparently selling the Medicare numbers of all Australians. (*Age*, 5 July 2017)

Finally, one of the worst-case biopolitical scenarios is to imagine that biometric data required for security reasons, like in facial recognition by means of iris scanners, could be hacked and used for criminal identity theft.

(20) It doesn't take a great stretch of the imagination to visualise what could happen <u>if that database</u> [*the FBI database*] <u>fell prey to a</u>

massive cyber attack like the ones we've witnessed in recent months. If at the time we happened to be relying on face recognition cameras to keep our phones, bank accounts, and perhaps even our cars, houses—hell, even our DNA—safe, then the game really would be on. (*Independent*, 23 July 2017)

2.5   STRATEGIES OF EXPERT DISCOURSE ABOUT BIG DATA

After retracing the main linguistic and discursive strategies in the news media's big data narrative, this section discusses a few highlights in the construction of expert discourse, focusing on big data coverage as an instance of expert opinion as a form of argumentation in science journalism (Goodwin and Honeycutt 2009; Greco Morasso and Morasso 2014). Built around the knowledge asymmetry between the expert and the lay person, expert opinion from the scientific community is a sensitive communication practice in the contemporary media ecosystem and, as such, requires that "the public should be involved in an argumentative partnership concerning the human, social and environmental implications of research" (Greco Morasso and Morasso 2014, p. 186). In her recent review of technology coverage, Sara M. Watson observes that "a wider circle of journalists, bloggers, and academics are contributing to a critical discourse about technology by contextualizing, historicizing, and giving readers tools for understanding our relationship to technologies in our everyday lives" (2016, p. 57).

To start with, partnership with the public involves easy accessibility to content. As the examples provided so far from the study corpus have already partially illustrated in the use of "data is the new ____" metaphor (Watson 2014), science and technology journalism routinely has recourse to discursive processes of reconceptualisation and recontextualisation of knowledge that aim to make the topics it deals with palatable and interesting for the publics. On the one hand, reconceptualisation "will first of all involve replacing a conceptual representation with one that is more suited to the intended addressee: simplification, explicitation, reformulation, comparison, metaphor and simile may all be involved" (Bondi et al. 2015, p. 8), as can be seen in examples 21–26.

(21) Companies are tracking more data about consumers than ever. Practically every click you make online creates new records in some distant database. (*Washington Post Blogs*, 7 January 2016)

(22) [R]obotics and artificial intelligence are two different, but related fields. Robots are machines that react to their environment—cars, for example, are not robotic since they respond to human instructions. (*Guardian*, 4 April 2016)

(23) "You go where the money is," said Steven Fazzari, a professor of economics at Washington University in St. Louis. "This is where companies are innovating and where there is demand". (*New York Times*, 24 April 2016)

(24) Privacy and the ownership and commercialization of the content any technology derives and how it's used are some of the subjects at stake. In short, once the information is out there, the genie is not going back in the bottle. (*USA Today*, 22 September 2016)

(25) Let's say there was a bank that locked the front door at night but left all its money sitting out on a table instead of securing it in the vault. That would be incredibly stupid—an invitation for thieves to break in and make off with the loot. And it's precisely what nearly every big company and government agency does with people's personal data. (*Los Angeles Times*, 29 September 2017)

(26) Saying technology is a threat to democracy is like saying food is bad for you because it causes obesity. (*Guardian*, 11 February 2018)

Recontextualisation "is aimed not only at making specialist knowledge available to the wider public as such, but also at making it relevant or interesting for the non-specialist reader, by highlighting the novelty and value of the area investigated, its relevance to the everyday life of readers/listeners, or to their communities and identities" (Bondi et al. 2015, p. 8). As such, it is related to the discursive construction of newsworthiness and the news values of Novelty, Impact and Superlativeness that have been investigated before (examples 27–29).

(27) The technology, called "the internet of things", is the way we will all live in the future, if tech companies are to be believed. (*Independent*, 5 May 2016)

(28) [*after purchasing LinkedIn*] Microsoft is betting LinkedIn, combined with data on how hundreds of millions of workers use its Office 365 online software, and consumer data from search behavior on Bing,

will "power a set of insights that we think is unprecedented," said James Phillips, vice president for business applications at Microsoft. (*New York Times*, 8 January 2017)

(29) Drawing on Keynesian parallels, Andy Haldane, Chief Economist for the Bank of England said in 2015 that a longer-term solution to countering the impact of automation would be to embrace "a world of progressively shorter working weeks, where mini-breaks become maxi-breaks". (*Age*, 4 May 2017)

The last research question that is addressed here concerns the identities of the main actors in the big data debate, their views and the ideologies thereby conveyed. We can retrace two forms of presence. In the first one, experts happen to be the authors of op-eds and columns dedicated to in-depth reporting on the issue, for example visionary scientist, academic and inventor Tim Berners-Lee, the founder of the world wide web, who comments on the recent examples of weaponisation of the web in the *Guardian*.

What is more frequently observable is to find expert opinion quoted within a news story as an authoritative source. The big names include economists like Carl Frey and Michael Osborne and philosopher Luciano Floridi from Oxford University, big data expert David Vaile from the Australian Privacy Foundation, US tech magazine *Wired*, doctors like Stanford cardiologist Alan Yeung, industry heavyweight gurus like Mark Zuckerberg,[23] chief executives, and politicians. Again, expert commentary is formulated in such a way as to sound accessible to the general public with the twofold aim of filling the knowledge gap and of representing technology and the changes it brings about as part of ordinary people's lives (in a few cases with clear business implications) (examples 30–32).

(30) [*on collecting health data by means of an app*] "This was a surprise," Yeung said. "A lot of people are spending most of their time sitting around—not even standing, not even going up and down". (*Washington Post*, 31 May 2016)

(31) "My personal challenge for 2016 is to build a simple AI to run my home and help me with my work," Mr Zuckerberg wrote at the beginning of the year. "You can think of it kind of like Jarvis in Iron Man". (*Independent*, 5 May 2016)

(32) The Australian Privacy Foundation's David Vaile said the government had "drunk the big data Kool Aid". "They seem to have set aside concerns about security, privacy, confidentiality and access controls," Mr. Vaile said. "Health is the most sensitive form of information, that can dog your entire life. It can affect insurance, family relationships, your capacity to get jobs or travel". (*Canberra Times*, 5 July 2017)

Discursively speaking, expert opinion is constructed as authoritative and reported in the text in order to advance knowledge claims, especially by means of epistemic stance markers (*there is no doubt, I think, simply put,* [she] *indicated that was possible*) and epistemic stance adverbials, introducing a subjective viewpoint and intensifying (*obviously*) or mitigating the proposition (*allegedly, perhaps*) (examples 33–38).

(33) Having technical skills for the future is <u>obviously</u> important—but they need to be complemented by cognitive functions, says Andrew Groth from Infosys, a technology services and consulting company. (*Australian Financial Review*, 2 December 2016)

(34) Professor Ariel Ezrachi, of Oxford University, who has written about companies' use of big data online, said: "<u>There is no doubt</u> the rules threaten these companies' core business model. That has to be weighed up against the interests of consumers, who would be empowered—it's a balancing act". (*Times*, 15 December 2016)

(35) One of the flaws in the current collection of big data, Mr. Glimcher [*a neuro-economist at New York University*] said, is that it misses large swaths of the population. <u>Simply put</u>, it skews to the wealthier and the healthier. (*New York Times*, 4 June 2017)

(36) Steven Mnuchin, the treasury secretary, has talked of taking action against Amazon because it <u>allegedly</u> does not pay its fair share of tax. (*Economist*, 12 August 2017)

(37) <u>Perhaps</u> the best way to think of the possible changes is to look for the parallels with previous industrial revolutions. (*Independent*, 23 January 2016)

(38) [*after the Equifax scandal*] Asked if the data breach signaled that new regulations might be warranted for companies handling sensitive data, White House Press Secretary Sarah Huckabee Sanders <u>indicated that</u>

was possible. "I think this is something we have to look into extensively," she said Monday. (*Los Angeles Times*, 12 September 2017)

In actual fact, given the unsettled nature of the big data debate, expert
opinion remains embedded in a highly polarised discursive field where
the main arguments ultimately "travel across the boundary between scientific and public contexts" (Goodwin and Honeycutt 2009, p. 28) to
reach the policy arena that is still in search of an answer.

2.6   Concluding Remarks

This chapter has focused on the big data debate in the global news media
to glean insights into social imaginaries in the handling of this highly
innovative socio-technical construct. After collecting a topic-specific,
hand-crafted study corpus tracing over a two-year period, the analysis
applied the dual methodology of CADS to elicit a number of empirical
findings conducive to the interpretation of larger stretches of text in context. The discourse analysis of the selected texts has shed further light
onto the main points of the storyline, the involved actors and the social
processes at work, once again showing the proactive role of discourse in
society. In fact, merit should be ascribed to CADS for discarding what
is ultimately an ideological division between quantitative and qualitative
methodologies to show instead that, when properly applied, their mutual
synergy contributes to the deeper understanding of the linguistic and discursive construction of meanings as an essential part of social practice,
without rejecting the researcher's intuition.

In our case, the positive narrative is undermined by the suspicion that
surveillance capitalism, now under the tyranny of algorithms, will be an
attack on democracy itself, in areas such as ethics, privacy, human rights,
journalism, intellectual property and economics. Big data appears to be
framed between two poles—data and information as opposed to rights
and privacy—whose gap has of late been emphasised by a number of data
scandals affecting business, health and politics, and culminating in the
major unforeseen event of Cambridge Analytica and Facebook.

The impact and fallacies of big data analytics undo any attempt to
circumscribe the topic to the realms of science and technology, showing instead the deep interconnectedness of technological innovation
with the world of business and finance, healthcare and politics. Since the
vulnerability of data seems to be equal to its value, this new resource is

surrounded by mounting cultural and political anxieties about its actual use that repeatedly surface in news discourse, for example by means of new compounds and figurative language that try to make sense of the phenomenon and translate it for the lay public. The contested nature of big data is also manifest in the unsolved argumentation regarding the topic across the corpus, which is left open-ended even when an appeal to expert opinion is made to bridge the distance between empowered stakeholders and ordinary citizens. Finally, ethical reflections with regard to this new technology have been imperfectly formulated, when not obfuscated, by the big data hype. This is the reason why the overview of the big data debate in the news media brings out an uneven landscape, with recent coverage showing a marked discontinuity since the end of March 2018 in the wake of the Cambridge Analytica data scandal. Arguably, the scandal managed to stimulate a commitment to data ethics in public discourse and not just within expert communities. Though disgracefully, the current public debate on data ethics was set in motion precisely by what the news media reported as a grossly unethical privacy breach and a disquieting example of the weaponisation of data by platforms.

Despite recent scandals and the grey areas yet to be clarified, the current technological debate resonates with the promise that the advantages of big data will ultimately override its drawbacks, first and foremost in the vital domain of healthcare.

NOTES

1. Over the weekend of 18 March 2018, the Sunday newspaper the *Observer*, which is owned by Guardian Media Group, revealed that Cambridge Analytica (CA), a data analytics firm that worked with Donald Trump's election campaign, had extracted Facebook data from 50 million user accounts. Between 2013 and 2015, Cambridge Analytica harvested profile data from millions of Facebook users, without those users' permission, and used that data to build a massive, targeted marketing database of each user's individual likes and interests. Using a personality profiling methodology, the company, which was financed by billionaire Republican donor Robert Mercer, backed by the then-future Trump White House adviser Steve Bannon and formed by high-powered rightwing investors, began offering its profiling system to dozens of political campaigns.
2. The affordances of LexisNexis have been of help in the selection, in particular the hit count and the preview window. It has been observed that

the focus on quality press and digital accessibility (especially LexisNexis) is a widespread practice in academic newspaper corpus design that is not exempt from the risk of allocating trust to the same upmarket news sources and of uncritically validating the ideology of mainstream media organs (Ali 2018). It disregards the polarised attitudes in volatile audiences worldwide that often result in "alienation from mainstream media" when "closely linked to perceived political bias" (Newman et al. 2018, p. 18). This behaviour, however, would seem to also hold true for the popular press, alternative news websites and social media. Since small sizeable corpora can be manually investigated, this was felt to be one of those cases in which the researcher's own critical insights could counterbalance the findings' circularity.

3. The *Australian* is published by Rupert Murdoch's NewsCorp. *Guardian Australia*, the *Guardian*'s third international digital edition launched in 2013, is owned by Guardian Media Group (GMG).

4. The *Guardian UK* has an entire section focused on big data, which includes a blog and a searchable database with information on all kinds of topics. The *New York Times* has a blog that is not specifically dedicated to big data, but deals heavily with data-related news items. The *LA Times* also has a Data Desk blog with reporters and programmers in the middle of the newsroom who collaborate on maps, databases, analysis and visualisation crafted from big data. Major news agencies are leading the way to a deep transformation in journalistic practices in which big data is not just a news topic but becomes a reporting tool.

5. "The concordance is defined more technically by Sinclair as 'a collection of the occurrences of a word-form, each in its own textual environment' [1991, p. 32]" (Partington et al. 2013, p. 17).

6. To remove stopwords, extremely common words that would be of little help in the analysis, such as conjunctions, prepositions, pronouns, possessives, auxiliary verbs *to be*, *to do* and *to have* in their different forms, and frequent abbreviations (*Co.*, *i.e.*, *Inc.*, *Ltd.*, *no.*, *pg.*), were excluded and a stopword list was built and uploaded to Sketch Engine.

7. *Data* occurs 2784 times in the corpus, 510 of which as *big data*, while the occurrences of *privacy* are 299 and those of *data protection* only 27.

8. The Quantified Self (QS) movement, or lifelogging, was founded by Gary Wolf and Kevin Kelly from *Wired* magazine in 2007 and has now spread worldwide. Its brand claim on the website is "self-knowledge through self-tracking". QS "is considered to be a movement to incorporate technology into data acquisition on aspects of a person's daily life in terms of inputs (e.g. food consumed, quality of surrounding air), states (e.g. mood, arousal, blood oxygen levels), and performance (mental and physical)" (Gurrin et al. 2014, p. 5).

9. The reference corpus was the English Web Corpus 2013 (enTenTen13), a corpus of general English made up of texts collected from the internet, amounting to over 19 billion words and preloaded to Sketch Engine.

10. "Generally, the higher value (100, 1000, ...) of Simple maths focuses on higher-frequency words (more common words), whereas the lower value (1, 0.1, ...) of Simple maths will rather prefer the words with lower frequency (more rare words)" (SketchEngine, n.d.).

11. There are 83 occurrences of *personal data* and 68 of *personal information*.

12. KWIC stands for Key Word in Context. In this format, the node word is placed in a central position with all lines vertically aligned around it.

13. Though newsworthiness is not only discursive but also material (events in their material reality) and cognitive (criteria internalised by journalists), this analysis investigates it as a textual feature (Caple and Bednarek 2016).

14. A word that follows the searched term with a high frequency is called its right collocate; a word that precedes it with a high frequency is called its left collocate.

15. Occurrences are as follows: *ethics* (14) (five of which in *data ethics*), *ethical* (26), *ethically* (1) and *ethicist* (2). *Unethical* is a *hapax legomenon*.

16. The concept of nudging is drawn from behavioural economics and describes designing choices in such a way that people will be guided by small suggestions and positive reinforcement to make the "right" choice unless they have strong alternative preferences (Puaschunder 2017).

17. The normalised frequency, or frequency per million, is used to compare frequencies between corpora of different sizes, as in this case.

18. In July 2017, a *Guardian Australia* journalist was able to purchase his Medicare details from a dark-web vendor on a popular auction site for illegal products at the cost of 0.0089 bitcoin, which is equivalent to US$22. Medicare card details can be used for identification fraud and "identification cards have been used by drug syndicates to buy goods and lease or buy property or cars" (*Guardian Australia*, 3 July 2017).

19. On 7 September 2017, Equifax, one of the three largest credit reporting providers in the US, revealed that from mid-May through July, the personal information of 145.5 million consumers was compromised. Exposed data included names, Social Security numbers, addresses, birth dates, and in some cases, driver's license numbers. An estimated 209,000 people also had their credit card information stolen.

20. Earlier in June 2018, PageUp, a multinational HR platform that manages job applications globally and is used by major Australian companies for their recruitment, revealed that it had been affected by a data breach.

21. Identity theft may happen through the stealing of biometric data, i.e. biometric signals "of body parts, such as finger, face, or iris" (Ratha et al. 2001, p. 628).

22. "All this comes as further confirmation of how we are turning the wondrous informational resource that is the internet into a cyber cesspit, where Russian trolls, Islamist terrorists, and corporate and political lobbyists peddle bile, propaganda, and lies" (*Financial Times*, 19 March 2018).

23. After the Cambridge Analytica scandal, as would be expected, Facebook CEO Mark Zuckerberg's largely uncontested stance as expert underwent a transformation, as he was called to appear on Capitol Hill, first before the joint Senate Judiciary on 10 April 2018 and, the following day, before the Congress House Energy and Commerce Committee. Zuckerberg was also interrogated by members of the European Parliament on 22 May 2018, but refused to testify before the UK Parliament. The extent to which his public persona was discursively challenged in these different contexts and controversies goes beyond the scope of this work, although several comments in the news media have been questioning Zuckerberg's assumption of full responsibility for Facebook's data abuse.

REFERENCES

Ajana, Bithaj. 2017. "Digital Health and the Biopolitics of the Quantified Self." *Digital Health* 3: 1–18. https://doi.org/10.1177/2055207616689509.

Ali, Samina. 2018. "Newspaper Corpus Design and Representativeness." WhatEvery1Says Project, 3 July. http://we1s.ucsb.edu.

Ames, Morgan G. 2018. "Deconstructing the Algorithmic Sublime." *Big Data & Society* 5, no. 1: 1–4. https://doi.org/10.1177/2053951718779194.

Baker, Paul. 2004. "Querying Keywords: Questions of Difference, Frequency, and Sense in Keywords Analysis." *Journal of English Linguistics* 32, no. 4: 346–359. https://doi.org/10.1177/0075424204269894.

Baker, Paul. 2006. *Using Corpora in Discourse Analysis*. London and New York: Continuum.

Baker, Paul, Costas Gabrielatos, Majid Khosravinik, Michał Krzyżanowski, Tony McEnery, and Ruth Wodak. 2008. "A Useful Methodological Synergy? Combining Critical Discourse Analysis and Corpus Linguistics to Examine Discourses of Refugees and Asylum Seekers in the UK Press." *Discourse & Society* 19, no. 3: 273–306. https://doi.org/10.1177/0957926508088962.

Baker, Paul, and Tony McEnery, eds. 2015. *Corpora and Discourse Studies: Integrating Discourse and Corpora*. Basingstoke and New York: Palgrave Macmillan.

Ball, Kirstie, Maria Laura Di Domenico, and Daniel Nunan. 2016. "Big Data Surveillance and the Body-Subject." *Body & Society* 22, no. 2: 58–81. https://doi.org/10.1177/1357034X15624973.

Bednarek, Monika, and Helen Caple. 2014. "Why Do News Values Matter? Towards a New Methodological Framework for Analysing News Discourse in Critical Discourse Analysis and Beyond." *Discourse & Society* 25, no. 2: 135–158. https://doi.org/10.1177/0957926513516041.

———. 2017. *The Discourse of News Values: How News Organizations Create 'Newsworthiness'.* Oxford: Oxford University Press.

Bondi, Marina, Silvia Cacchiani, and Davide Mazzi, eds. 2015. *Discourse In and Through the Media: Recontextualizing and Reconceptualizing Expert Discourse.* Newcastle upon Tyne: Cambridge Scholars Publishing.

Caple, Helen, and Monika Bednarek. 2016. "Rethinking News Values: What a Discursive Approach Can Tell Us about the Construction of News Discourse and News Photography." *Journalism* 17, no. 4: 435–455. https://doi.org/10.1177/1464884914568078.

Caulfield, Timothy. 2004. "Biotechnology and the Popular Press: Hype and the Selling of Science." *Trends in Biotechnology* 22, no. 7: 337–339. https://doi.org/10.1016/j.tibtech.2004.03.014.

Economist. 2014. "Self-Made Wealth in America: Robber Barons and Silicon Sultans." 30 December. https://www.economist.com/briefing/2014/12/30/robber-barons-and-silicon-sultans.

Floridi, Luciano. 2011. *The Philosophy of Information.* Oxford: Oxford University Press.

Galtung, Johan, and Mari Holmboe Ruge. 1965. "The Structure of Foreign News: The Presentation of the Congo, Cuba and Cyprus Crises in Four Norwegian Newspapers." *Journal of Peace Research* 2, no. 1: 64–91.

Garzone, Giuliana, and Francesca Santulli. 2004. "What Can Corpus Linguistics Do for Critical Discourse Analysis?" In *Corpora and Discourse*, edited by Alan Partington, John Morley, and Louann Haarman, 351–368. Bern: Peter Lang.

Goodwin, Jean, and Lee Honeycutt. 2009. "When Science Goes Public: From Technical Arguments to Appeals to Authority." *Studies in Communication Sciences* 9, no. 2: 19–30.

Graves, Christopher, and Sandra Matz. 2018. "What Marketers Should Know about Personality-Based Marketing." *Harvard Business Review*, May 2. https://hbr.org/2018/05/what-marketers-should-know-about-personality-based-marketing.

Greco Morasso, Sara, and Carlo Morasso. 2014. "Argumentation from Expert Opinion in Science Journalism: The Case of Eureka's Fight Club." In *Rhétorique et cognition - Rhetoric and Cognition: Perspectives théoriques et stratégies persuasives - Theoretical Perspectives and Persuasive Strategies*, edited by Thierry Herman and Steve Oswald, 185–213. Bern: Peter Lang.

Gurrin, Cathal, Alan Smeaton, and Aiden R. Doherty. 2014. "LifeLogging: Personal Big Data." *Foundations and Trends® in Information Retrieval* 8, no. 1: 1–107. http://dx.doi.org/10.1561/1500000033.

Kilgarriff, Adam. 2009. "Simple Maths for Keywords." *Proceedings of the Corpus Linguistics Conference CL2009*, edited by Michaela Mahlberg, Victorina González-Díaz, and Catherine Smith, article number 171, 1–6. Liverpool: University of Liverpool. http://ucrel.lancs.ac.uk/publications/cl2009.

———. 2012. "Getting to Know Your Corpus." In *Text, Speech and Dialogue*. Lecture Notes in Computer Science, vol. 7499, edited by Petr Sojka, Aleš Horák, Ivan Kopeček, and Karel Pala, 3–15. Heidelberg and Berlin: Springer. https://doi.org/10.1007/978-3-642-32790-2_1.

Kilgarriff, Adam, and Gregory Grefenstette. 2003. "Introduction to the Special Issue on the Web as Corpus." *Computational Linguistics* 29, no. 3: 333–347. https://doi.org/10.1162/089120103322711569.

Kilgarriff, Adam, Vít Baisa, Jan Bušta, Miloš Jakubíček, Vojtěch Kovář, Jan Michelfeit, Pavel Rychlý, and Vít Suchomel. 2014. "The Sketch Engine: Ten Years On." *Lexicography* 1, no. 7: 7–36. https://doi.org/10.1007/s40607-014-0009-9.

Kitchin, Rob. 2014. *The Data Revolution: Big Data, Open Data, Data Infrastructures & Their Consequences.* London: Sage.

Koester, Almut. 2010. "Building Small Specialised Corpora." In *The Routledge Handbook of Corpus Linguistics*, edited by Anne O'Keeffe and Michael McCarthy, 66–79. Abingdon and New York: Routledge.

Koteyko, Nelya. 2010. "Mining the Internet for Linguistic and Social Data: An Analysis of Carbon Compounds in Web Feeds." *Discourse & Society* 21, no. 6: 655–674. https://doi.org/10.1177/0957926510381220.

Lanzing, Marjolein. 2016. "The Transparent Self." *Ethics and Information Technology* 18, no. 1: 9–16. https://doi.org/10.1007/s10676-016-9396-y.

Lohr, Steve. 2012. "How Big Data Became So Big." *New York Times*, 11 August. http://www.nytimes.com/2012/08/12/business/how-big-data-became-so-big-unboxed.html.

Metcalf, Jacob, and Kate Crawford. 2016. "Where Are Human Subjects in Big Data Research? The Emerging Ethics Divide." *Big Data & Society* 3, no. 1: 1–14. https://doi.org/10.1177/2053951716650211.

Mittelstadt, Brent Daniel, Patrick Allo, Mariarosa Taddeo, Sandra Wachter, and Luciano Floridi. 2016. "The Ethics of Algorithms: Mapping the Debate." *Big Data & Society* 3, no. 2: 1–21. https://doi.org/10.1177/2053951716679679.

Newman, Nic. 2018. *Journalism, Media, and Technology Trends and Predictions 2018.* Oxford: Reuters Institute for the Study of Journalism, The University of Oxford.

Newman, Nic, Richard Fletcher, Antonis Kalogeropoulos, David A. L. Levy, and Rasmus Kleis Nielsen. 2018. *Reuters Institute Digital News Report 2018*, 14 June. Oxford: Reuters Institute for the Study of Journalism, The University of Oxford. https://ssrn.com/abstract=3245355.

C' Halloran, Kieran. 2010. "How to Use Corpus Linguistics in the Study of Media Discourse." In *The Routledge Handbook of Corpus Linguistics*, edited by Anne O'Keeffe and Michael McCarthy, 563–577. Abingdon and New York: Routledge.

Oxford Internet Institute. 2017. "Digital Ethics Lab." https://www.oii.ox.ac.uk.

Partington, Alan. 2004a. "Corpora and Discourse, A Most Congruous Beast." In *Corpora and Discourse*, edited by Alan Partington, John Morley, and Louann Haarman, 11–20. Bern: Peter Lang.

———. 2004b. "'Utterly Content in Each Other's Company': Semantic Prosody and Semantic Preference." *International Journal of Corpus Linguistics* 9, no. 1: 131–156. https://doi.org/10.1075/ijcl.9.1.07par.

Partington, Alan, Alison Duguid, and Charlotte Taylor. 2013. *Patterns and Meanings in Discourse: Theory and Practice in Corpus-Assisted Discourse Studies (CADS)*. Amsterdam and Philadelphia: John Benjamins.

Perelman, Chaïm, and Lucie Olbrechts-Tyteca. 1969. *The New Rhetoric: A Treatise on Argumentation*. Notre Dame, IN: University of Notre Dame Press.

Portmess, Lisa, and Sara Tower. 2015. "Data Barns, Ambient Intelligence and Cloud Computing: The Tacit Epistemology and Linguistic Representation of Big Data." *Ethics and Information Technology* 17, no. 1: 1–9. https://doi.org/10.1007/s10676-014-9357-2.

Puaschunder, Julia M. 2017. "The Nudging Divide in the Digital Big Data Era." *International Journal of Research in Business, Economics and Management* 4: 11–12, 49–53. https://ssrn.com/abstract=3007085.

Puschmann, Cornelius, and Jean Burgess. 2014. "Big Data, Big Questions| Metaphors of Big Data." *International Journal of Communication* 8: 1690–1709. http://ijoc.org/index.php/ijoc/article/view/2169.

Quantified Self Institute. n.d. "What Is Quantified Self?" https://qsinstitute.com.

Ratha, Nalini K., Jonathan H. Connell, and Ruud M. Bolle. 2001. "Enhancing Security and Privacy in Biometrics-Based Authentication Systems." *IBM Systems Journal* 40, no. 3: 614–643. https://ieeexplore.ieee.org/document/5386935.

Schofield, Alexandra, Laure Thompson, and David Mimno. 2017. "Quantifying the Effects of Text Duplication on Semantic Models." In *Proceedings of the 2017 Conference on Empirical Methods in Natural Language Processing*, edited by Martha Palmer, Rebecca Hwa, and Sebastian Riedel, 2737–2747. Copenhagen: Association for Computational Linguistics. https://aclweb.org/anthology/D17-1290.

Scott, Mike. 1997. "PC Analysis of Key Words—And Key Key Words." *System* 25, no. 2: 233–245. https://doi.org/10.1016/S0346-251X(97)00011-0.

———. 1999. *WordSmith Tools Help Manual*. Version 3.0. Oxford: Mike Scott and Oxford University Press.

———. 2010. "Problems in Investigating Keyness, or Cleaning the Undergrowth and Marking Out Trails..." In *Keyness in Texts*, edited by Marina Bondi and Mike Scott, 43–57. Bern: Peter Lang.

Sinclair, John. 1991. *Corpus, Concordance, Collocation*. Oxford: Oxford University Press.

Sketch Engine. n.d. "Simple Maths." https://www.sketchengine.eu/documentation/simple-maths.

Stubbs, Michael. 1996. *Text and Corpus Linguistics: Computer-Assisted Studies of Language and Culture*. Oxford: Blackwell.

———. 2001. *Words and Phrases: Corpus Studies of Lexical Semantics*. Oxford: Blackwell.

Thornbury, Scott. 2010. "What Can a Corpus Tell Us about. Discourse?" In *The Routledge Handbook of Corpus Linguistics*, edited by Anne O'Keeffe and Michael McCarthy, 270–287. Abingdon and New York: Routledge.

van Dijk, Teun A. 1988. *News as Discourse*. Hillsdale, NJ: Lawrence Erlbaum Associates.

Watson, Sara M. 2014. "Data Is the New '____': Sara M. Watson on the Industrial Metaphor of Big Data." *DIS Magazine*. http://dismagazine.com/discussion/73298/sara-m-watson-metaphors-of-big-data.

———. 2016. "Toward a Constructive Technology Criticism." Tow Center for Digital Journalism White Papers. New York: Columbia University. https://doi.org/10.7916/D86401Z7.

CHAPTER 3

Big Data and Healthcare

Abstract The chapter analyses a corpus of Biomedical Big Data in the news media in order to elicit the linguistic and discursive strategies that frame the big data and healthcare debate in the news media and in popularised scientific discourse. A number of exploratory probes to retrieve keywords are followed by the discourse analysis of wider stretches of text and a closer focus on the representation of agency by means of periphrastic causative verbs. Two main actors—patients and doctors—are discursively foregrounded, with technology and information as helpers that increasingly mediate between them, reflecting the shift towards self-management, from institutional to patient-centred care. Findings show that innovation in healthcare is represented as a challenge and a unique opportunity that includes predictive, precision and patient-centric care.

Keywords Big data · Healthcare discourse · Patient-centredness · Precision care · Predictive analytics

3.1 STUDY BACKGROUND

"New technologies have launched the life sciences into the age of big data" and, in the words of Eric D. Green, director of the National Human Genome Research Institute in Bethesda, "the life sciences are becoming a big data enterprise" (Singer 2013, n.p.). The health-related internet of things (H-IoT)—internet-enabled devices for monitoring

© The Author(s) 2019
M. C. Paganoni, *Framing Big Data*,
https://doi.org/10.1007/978-3-030-16788-2_3

and managing the health and well-being of users outside of traditional medical institutions—has rapidly spread to support healthcare. Over the last fifteen years, governments in advanced economies "have all made long-term, multibillion dollar investments in health information technologies, including electronic health records" (Sittig and Singh 2012, p. 1479) that contain a host of clinical data. Modern healthcare systems routinely produce an abundance of electronically stored data on an ongoing basis. The use of data has great potential in "research, regulation, policymaking, surveillance, and clinical care" (Lipworth et al. 2017, p. 489) and in both of the major challenges of healthcare: to improve patient experience and health outcomes and to curb the rising costs of public health.

As for the first one, i.e. better care, H-IoT promises a number of benefits, paving the way for more personalised diagnosis, with healthcare evolving towards a system of predictive, preventive, and precision care. Traditionally confined to the realm of genetics, *predictive* medicine will soon be able to calculate individual risk for a variety of healthcare outcomes and determine optimal, personalised treatment options. Combined information about patient populations, behaviours and outcomes will give providers crucial insights for designing and executing *preventive* care strategies and targeted patient outreach and education. Finally, *precision* medicine approaches are enabled by data leveraged from direct and indirect sources to provide a more holistic view of an individual patient.

Second, as regards social spending, H-IoT technologies increasingly play a key role in health management for purposes including disease prevention, real-time tele-monitoring of patients' functions, testing of treatments, fitness and well-being monitoring, medication dispensation, and health research data collection. Big data analytics is also helping the public sector to deliver more effective and efficient services and produce positive outcomes that improve the quality of people's lives while cutting costs. However, big data analytics "also raises a host of ethical problems stemming from the inherent risks of Internet-enabled devices, the sensitivity of health-related data and their impact on the delivery of healthcare" (Mittelstadt 2017a, p. 157). Because of technology, the healthcare industry is not only more efficient but also more vulnerable to cyberattacks.

After reviewing the big data debate in the previous chapter and in light of the specific needs and requirements of healthcare, the present chapter intends to investigate how big data analytics in healthcare is framed by the news media, emphasising the role they play in knowledge dissemination. In order to do so, it will address the following research questions:

- What are the recurrent linguistic and discursive strategies employed in the news media to frame the big data debate in healthcare settings?
- If compared with the main highlights of big data in news discourse, are there perceivable differences?
- What (bio)ethical issues are mentioned?

3.2   MATERIALS AND METHODS

This part of the analysis of news discourse is characterised by a thematic focalisation on a subtopic of the big data debate, the use of big data in the biomedical field and healthcare. To this purpose, a specific corpus was compiled and named "Biomedical Big Data" (BBD). The BBD corpus comprises 101 items published over the time span of two years, from January 2016 to March 2018, and counting over 107,660 words. Arguably, the 11.3 type-token ratio (TTR) of the BBD corpus, which is higher than the 7.65 TTR of the Big Data corpus, seems to roughly indicate greater lexical diversity and complexity (Baker 2006, p. 52) in the popularisation of biomedical research and public health discourse.

The same pool of global online news sources selected for the exploratory overview of the big data topic in Chapter 2 was searched, with just three further additions.[1] The main concept of the collected news items was the use of big data in the biomedical healthcare sector, deliberately avoiding passing references to tech innovation. Again, priority was given to digital accessibility of the items that were downloaded from LexisNexis Academic, Europresse and Google News.

The majority of news items were collected from the mainstream quality press in the UK, the US and Australia. Additional articles were also retrieved from science communication magazines, like *Australian Popular Science*, *HealthITAnalytics* and *Quanta Magazine*, technology magazines like *Wired* and business magazines like *Forbes* and *Fortune*, whenever the contributions were seen as meaningful instances of an in-depth treatment of big data in healthcare addressed to the general public. To be noted, nine sponsored content articles, published in the Partnerships in Practice section of *Guardian UK* and paid by the Brother Group, were included on the basis of their chosen topics.[2] As with the Big Data corpus, the BBD corpus is likewise topic-specific and user-defined (Bednarek and Caple 2014, p. 137), with comprehensive news coverage over a two-year span. Size-wise, it lends itself to qualitative analysis, since it is possible to manually check each single item and not just a sample, as is the case with large corpora (Koester 2010). All these empirical criteria are preset by the researcher to guarantee the degree of representativeness required for the analysis (Kilgarriff and Grefenstette 2003).

The decision to investigate biomedical big data separately was based upon the researcher's perception of a different discursive frame at work in the case of healthcare, a hypothesis that was progressively formulated during the months in which the larger Big Data corpus was compiled. The need was felt, however, to ground such intuition in a more articulated discussion of empirical findings before attempting an explanation.

In line with the analysis that was carried out in the preceding chapter, the selected methodology is Corpus-Assisted Discourse Studies (CADS), with the latest version of Sketch Engine (Kilgarriff et al. 2014) as the software tool for Corpus Linguistics. This mixed-methods approach involves the deployment of quantitative and qualitative procedures whereby findings retrieved by Sketch Engine (keywords, concordances and collocations) are progressively expanded to longer stretches of text whose interpretation may shed light on culture and society (Garzone and Santulli 2004; Partington 2004; Baker 2006; Baker et al. 2008; Partington et al. 2013; Baker and McEnery 2015). Situated as it is "in the midst of constantly emerging organizational discourses and practices" (Iedema and Carroll 2014, p. 189), Discourse Analysis has been reputed a flexible approach to the imperative of innovation in healthcare.

In order to anchor the discourse analysis of the corpus to insights into the use of big data in medicine and public healthcare settings, expert advice as to what is an authoritative source for healthcare professionals was directly sought from researchers in the biomedical field. A few recent pieces on big data (Obermeyer and Emanuel 2016; Chen and Asch 2017; Haug 2017; Obermeyer and Lee 2017; Char et al. 2018) in the Perspective section of the *New England Journal of Medicine*, one of the world's top medical journals, were recommended. These views were further integrated with bibliographical suggestions from the rest of the critical literature and with online searches. The aim was to get a better grasp of how the medical community responds to big data and develops an informed opinion. A trend in contemporary medical journals, perspectives are opinion pieces that discuss challenging and controversial questions relating to medicine and health in ways that are accessible to people from all walks of life who wish to become more knowledgeable in the medical and healthcare fields.

3.3   BIG DATA IN THE BIOLOGICAL AND BIOMEDICAL DOMAINS AND HEALTHCARE SETTING

Health is one of the areas in which the internet of things promises great social benefits. Big data and data science have developed numerous machine learning-based techniques and technologies that are applied to improve human health by solving unprecedented computational challenges. Big data analytics allows health providers to save costs through optimal management of resources.

IoT health makes technology businesses thrive. Smartphone apps that range "from pedometers to measure how far you walk in a day, to calorie counters to help you plan your diet" have been followed by "a steady stream of dedicated wearable devices" (Marr 2015, n.p.), such as fitness watches and Google Glass. Wearable and miniaturised sensor devices for self-tracking and lifelogging practices generate real-time flows of data that measure physiological data like heart rate, blood pressure and glucose levels, as well as health parameters having to do with exercise, sleep, nutrition and lifestyle (Gurrin et al. 2014).

It comes as no surprise, then, that big data should be hailed as highly innovative and transformative for healthcare, where it is expected to foster major advances. As has been seen, the news media tend to cover big data across the whole spectrum—from business to politics—in a polarised way, especially on the spur of data leaks and recent frauds and scandals. Despite several ethical concerns, instead, big data in the biomedical and healthcare setting is represented as fully unlocking its potential for positive change and therefore is not so fraught with contention.

The top content keywords and bigrams, ordered by keyness, were automatically extracted by Sketch Engine in order to outline the "aboutness" of the corpus (Scott 1997, 1999), i.e. its outstanding concepts. The keyness score measures the higher frequency of a word or phrase in the study corpus compared to a reference corpus, in this case, the enTenTen13 corpus of general English.[3] The score was computed using Sketch Engine's statistical formula, described by its founder as "simple maths for keywords" (Kilgarriff 2009, p. 1). As with the Big Data corpus, the N variable in this formula, or "simplemaths parameter" (Kilgarriff 2012, p. 5), was manually set for mid-frequency words ($N = 500$, i.e. from rank 500 upwards) in order to exclude rare words.[4] To limit redundancy, proper nouns like NHS and Medicare Australia that are included in the concept of *healthcare system* were not considered, nor were *dark web* (in the Medicare Australia scandal) and DeepMind, the British artificial intelligence company owned by Alphabet Inc., Google's parent company, since (almost inevitably!) Google is already in the list. The manually edited list below (Table 3.1) contains twenty-one key content lemmas and bigrams out of the top fifty, showing their keyness score and raw frequency.

As was to be expected, the semantic fields of *health* and *healthcare* are prioritised. In light of the previous discussion on big data in news discourse, what appears counterintuitive, instead, is the noticeable de-emphasis of *privacy*, with a total of sixty-five hits, which explains the discretionary cut-off at that point. A possible explanation is the common assumption in bioethical thinking that privacy may be sacrificed for the sake of a greater good such as public health.

> In existing bioethical scholarship, the challenges of big data are most commonly framed as reflecting the need to balance individual research participant rights against the common good; and principles such as individual autonomy and respect for persons against potentially competing principles such as altruism and solidarity. (Lipworth et al. 2017, p. 491)

In the news media, however, this ethical dilemma is occasionally simplified as a conflict between the common good and individual privacy in which the latter is always the loser. Instead of problematising the complexity that technological innovation brings about, this stance is premised on the bias that technology is inevitable and inexorable, a force that reshapes society "reconfiguring it in accordance with its internal workings" (Fisher 2010, p. 17), as in example 1.

(1) "There are tremendous advantages to big data in healthcare," said Gerard Magill, a professor of healthcare ethics at Duquesne University in Pittsburgh. "It's about creating a comprehensive approach to using medical information." The trade-off: Say goodbye to individual

Table 3.1 Keywords and bigrams (by keyness) in the Biomedical Big Data corpus

No.	Keywords (BBD)	Keyness	Frequency	Bigrams (BBD)	Keyness	Frequency
1	data	10.20	905	artificial intelligence	1.89	56
2	patient	6.63	535	health care	1.55	44
3	health	5.71	559	heart disease	1.45	30
4	healthcare	4.85	264	machine learning	1.45	28
5	technology	3.77	284	precision medicine	1.42	26
6	medical	3.54	237	chief executive	1.33	21
7	hospital	3.07	169	health information	1.28	18
8	doctor	3.00	174	patient care	1.22	14
9	disease	2.97	171	population health	1.19	12
10	care	2.86	239	deep learning	1.18	11
11	record	2.53	165	chronic disease	1.17	11
12	information	2.50	259	healthcare system	1.17	11
13	access	2.36	147	health insurance	1.17	13
14	app	2.34	110	value-based care	1.16	10
15	insurer	2.31	85	health record	1.16	10
16	Google	2.16	102	personal information	1.15	13
17	risk	2.15	124	health management	1.14	9
18	intelligence	2.08	77	preventive care	1.14	9
19	AI (artificial intelligence)	2.08	70	public health	1.13	10
20	machine	2.01	92	wearable technology	1.13	8
21	privacy	1.91	65	personal health	1.13	8

privacy. "Big data requires that information; it's nonnegotiable," Magill said. "Individual privacy is gone for the common good". (*Los Angeles Times*, 11 October 2016)

The relatively low keyness of *privacy* in the BBD corpus does not elide its being predominantly associated with a negative semantic prosody, which

Table 3.2   Edited concordance lines of *privacy*

seeking to address concerns about	*privacy*	and coercion
the accompanying threats to patients'	*privacy*	Within a few years databases of millions
Those who fear an invasion of	*privacy*	and choose not to buy these policies
medical data relating to 1.6 m patients.	*Privacy*	advocates see this as "worrying"
security, trust and	*privacy*	are compromised.
the latest data security and	*privacy*	scandal to rock the Australian government
there are	*privacy*	concerns about collecting data
systemic problems of accessibility and	*privacy*	when using medical data
However, for	*privacy*	campaigners it's part of a "Big Brother" moment
for government agencies to notify the	*privacy*	commissioner of certain types of data breaches

results from collocation with words having strong negative semantic characteristics, both to the right and to the left (Table 3.2). The textual and discoursal relationships of *privacy* with notions of fear, threat and surveillance ("Big Brother") convey the public's perception of manipulation and loss of intimate and personal information, which is the major downside in big data analytics and has not been properly addressed yet.

This line of reasoning can be further extended by interpreting the one hundred and two hits of *Google*. As a matter of fact, mentions of the Mountain View conglomerate should be integrated with fourteen additional hits of *Alphabet* and ninety-nine hits of *DeepMind*, since they belong to same corporate structure. Alphabet, the holding company founded in 2015 whose largest wholly owned subsidiary is Google, also owns London-based DeepMind Technologies Limited, one of the labs that aspire to pioneer "data-driven tools and techniques, particularly machine learning methods that underpin artificial intelligence [...] in improving healthcare systems and services" (Powles and Hodson 2017, p. 351).

In the BBD corpus, *Google* is characterised by a mixed semantic prosody that alternates between factual statements and positive semantic associations on the one hand, and negative semantic associations on the other, as is illustrated in Table 3.3.

In 2015, DeepMind made the first data-sharing agreement with the UK's National Health Service (NHS) until September 2017, signing a

Table 3.3   Edited concordance lines of *Google*

X-ray scans reviewed by radiologists.	*Google*	researchers used similar machine-learning methods
By looking at the human eye,	*Google's*	algorithms were able to predict whether
Any patient can use	*Google*	to find out more about the symptoms
widely used today in systems such as	*Google's*	image search and Facebook
in a study lead-authored by	*Google*	scientists published in JAMA, predicted diabetic eye disease
for tracking fitness, smart watches and	*Google*	glasses, the biggest sectors to be impacted include healthcare
	***	
Not just because	*Google*	, a sprawling octopus of a company with tentacles in all
better than your closest friend.	*Google*	knows your thoughts, and even completes your sentences.
the potential risks of	*Google*	dominance over a new critical technology for the NHS
the potential to give DeepMind, and thus	*Google*	too much power over the NHS
drawbacks of big data. Recent news that	*Google*	received access to the healthcare data of up to 1.6 million patients

deal with the Royal Free NHS Trust and the three hospitals it operates, where an estimated 1.6 million patients are treated annually. According to the agreement, DeepMind gained "access to all admissions, discharge and transfer data, accident and emergency, pathology and radiology, and critical care at these hospitals" and the "historical medical records data on patients who [*had*] been treated at the hospitals" (Lomas 2016, n.p.). However, in July 2017 after a year-long investigation, the UK's Information Commissioner's Office (ICO)[5] ruled that the NHS had violated data protection rules and asked for correctives in future trials (Burgess 2017, n.p.). In a blog post published on the ICO website after the report, Elizabeth Denham, the Information Commissioner, made such a point, arguing not against the use of big data in healthcare but against the rhetoric of technology as an uncontrollable process (example 2).

(2) **It's not a choice between privacy or innovation**
It's welcome that the trial looks to have been positive. The Trust has reported successful outcomes. Some may reflect that data protection rights are a small price to pay for this. But what stood out to me on

looking through the results of the investigation is that the shortcomings we found were avoidable. <u>The price of innovation didn't need to be the erosion of legally ensured fundamental privacy rights</u>. I've every confidence the Trust can comply with the changes we've asked for and still continue its valuable work. (Denham 2017, n.p.)

Another aspect that deserves critical scrutiny is the higher rank of *patient(s)* (with 535 hits) compared to *doctor(s)* (174). Though the gap is reduced by adding synonyms like *clinician(s)* (44), *caregiver(s)* (3) and *health/healthcare/medical provider(s)* (26), the imbalance is evident. Only ten binomials altogether—*doctors and patients, patients and clinicians, patients and caregivers, patients and medics, a doctor and a patient*—place the two main participants in healthcare on a par.

What this finding indicates is undoubtedly the priority that should always be given to patients, the moral standing that goes under the name of "patient-centred care",[6] a phrase that is common vernacular in healthcare settings (example 3).

(3) A poll in February by PwC concluded that "people are demanding more convenient care, more access and a bigger say in decisions about them and their care… They're willing to have their care in non-traditional settings, from non-traditional players". (*Financial Times*, 1 April 2016)

A look into concordances and collocations, however, reveals new additional elements in the direction of "remote patient monitoring and robotic surgery" (*Times*, 24 June 2017), algorithms and machine learning in patient care, and "patient activation", all practices that are introduced or boosted by big data (Fiebig 2017) and debated in the news media and public discourse (examples 4–7).

(4) [T]he advance of digital technology [...] is increasingly playing a role in <u>how patients manage their conditions and companies communicate the benefits of their medicines to doctors</u>. (*Financial Times*, 11 July 2017)

(5) A medical device using <u>deep learning</u> to analyze long-term neural data could effectively predict seizures in patients with epilepsy and reduce their disease burden, according to a study published in *eBioMedicine*. (*HealthItAnalytics*, 1 February 2018)

(6) "The biggest challenge in healthcare is about making sure people take responsibility for their own health," says Mr Blinder [*CEO of Tictrac, a lifestyle platform*]. "<u>A lot of us feel somewhat entitled to healthcare</u>, that someone else will take care of us." […] If what Mr Blinder is saying sounds familiar, then that's probably because we've heard it before. The NHS calls it "*patient activation*"—a buzzword for their five-year plan <u>to help people take better care of their own health and alleviate the burden on the health service</u>. (*Independent*, 30 May 2017)

(7) **Patient activation**
People's ability to manage their own health and wellbeing
"<u>Patient activation</u>" describes the knowledge, skills and confidence a person has in managing their own health and care. Evidence shows that when people are supported to become more activated, they benefit from better health outcomes, improved experiences of care and fewer unplanned care admissions. (*NHS England*, n.d.)

As to this last point, if we support the claim that "agency based on language use and language structure […] can be integrated with a social theory of agency" (Duranti 2004, p. 452), then the linguistic representation of agency and of participants in agentive relationships may also give a sense of the direction in which healthcare is heading. In particular, the focus has been placed on periphrastic causative constructions (Wolff and Song 2003), i.e. grammatical constructions that express a causal relation in which the occurrence of a certain effect is entailed.[7] In the first instance, the Agent is the Subject of an active clause with a periphrastic causative that does not embed an Object/Patient, as in examples 8–12, in which *help* is followed by the infinitive. To be noted, the Agent can be a non-sentient being of the H-IoT area (*wearables, devices, technology*).

(8) Wearables can also <u>help manage</u> the symptoms of illnesses. (*Guardian*, 18 May 2016)

(9) The research might in future, for example, <u>help determine</u> the optimal lay-off time for a head-injured rugby player. (*Age*, 1 January 2017)

(10) The experiment determined if basic reminder devices could <u>help boost</u> people's medication compliance. (*Australian Popular Science*, 5 May 2017)

(11) The global mantra is that we must move towards value-based healthcare in which three Ps are critical: prevention, precision and personalisation. Technology can <u>help address</u> all three challenges, particularly through the smart use of big data and artificial intelligence. (*Financial Times*, 15 May 2017)

(12) His firm could ultimately map the genomes of 50 million cancer patients alone a year, <u>helping identify</u> the genetic traits that contribute to the disease. (*Telegraph*, 20 September 2017)

In the second instance, we find a transitive sentence in the active voice with a periphrastic causative that embeds a Patient upon which the action impinges. In examples 13–19, the Agent is some kind of technology which is thus granted grammatical agency and seen as an enabler. The semantic role of the Patient is covered by the medical category, patients themselves and consumers, with the only exception of "data speaking for themselves" in example 20.

(13) Genomics technology can <u>help doctors to diagnose</u> genetic diseases more quickly. (*Times*, 20 June 2016)

(14) An evolving ecosystem of connected health technologies such as wearables, telehealth, artificial intelligence and virtual reality will <u>empower consumers to be</u> the new king or queen of their own healthcare decision-making process. (*Forbes*, 8 March 2017)

(15) The trust is testing wearable technology <u>to let patients carry out</u> walk tests at home [...]. The hope is both to save the NHS money, <u>enable patients to stay</u> in their own home for longer and <u>help doctors identify, and react to,</u> problems more quickly. (*Guardian*, 31 March 2017)

(16) A computer's ability to spot anomalies in data <u>can help clinicians identify</u> unusual signs and symptoms in patients better. (*Independent*, 10 April 2017)

(17) Tictrac's core function – <u>helping people monitor and improve health habits</u> through data tracking – is replicated by countless other apps. (*Independent*, 30 May 2017)

(18) Within a few years databases of millions of patients' DNA data or clinical information are expected to play a much bigger role in <u>helping clinicians diagnose</u> disease. (*Telegraph*, 17 September 2017)

(19) Microsoft developed a platform <u>to enable medical centers</u> to create virtual assistants for patients. (*New York Times*, 26 December 2017)

(20) With machine learning, according to a September editorial in *The New England Journal of Medicine*, "we <u>let the data speak</u> for themselves". (*Canberra Times*, 1 January 2017)

In a nutshell, the picture that emerges from the analysis of agency would seem to confirm the ongoing shift towards H-IoT supported self-management and from institutional to person-centred care, with the aim of achieving better outcomes and lower costs.

3.3.1   Advantages and Disadvantages of Big Data in Healthcare

The investigation of lexical and discursive features in the BBD corpus has provided a number of insights that describe what are felt to be advantages and disadvantages in the use of big data in the healthcare setting.

The datification of health data and the adoption of Electronic Health Records mean that everyone has a "digital care footprint" (*USA Today*, 2 March 2016). Data-driven medicine results in a personalised predictive cloud for each patient and produces actionable guidelines and insights in public health. Precision medicine can take advantage of state-of-the-art genomic sequencing. Machine learning appropriates the workings of the human brain with amazing health outcomes such as the enhancement of "nervous system recovery through neurobiologics, neural interface training, and neurorehabilitation" (Krucoff et al. 2016) and the applications of neurobionics and robotic assistive technology[8] to repair and replace impaired functions of the nervous system. Still, there are "many arenas where humans beat computers":

(21) "Today's machine learning fails where humans excel," said Jacob Vogelstein, who heads the program at the Intelligence Advanced Research Projects Activity (IARPA). "<u>We want to revolutionize machine learning by reverse engineering the algorithms and computations of the brain</u>". (*Quanta Magazine*, 6 April 2016)

H-IoT can capture data for a variety of purposes, from remote patient monitoring to controlling non-compliance with medication. Lifelogging is often associated with healthcare self-monitoring and based around some form of personalised healthcare or wellness (Lanzing 2016).

It streams valuable patient-generated data flows. Still in its infancy, "patient-initiated" research may prove to be the new gold standard for the study of uncommon medical conditions. Technology, it would seem, favours patient-centred care (Kreindler 2015).

By contrast, keeping health data separate and therefore not interoperable creates a disconnect between clinicians and data scientists, increases risk and impedes research (examples 22–24). This deadlock can be overcome by the practice of greater *data* and *information sharing* (18 and 7 occurrences respectively) that is recommended in example 25 to untrap restricted data.

(22) [Australia] has been notoriously slow at adopting technology despite a high standard of living and education. (*Sydney Morning Herald*, 29 July 2016)

(23) One of the biggest challenges is to free data from the silos in which it too often remains, an issue that affects both patient care and medical research. (*Fortune*, 1 September 2017)

(24) Although health records are increasingly electronic, they are often still trapped in silos. (*Economist*, 1 February 2018)

(25) Dr Jurgi Camblong, founder and chief executive of Sophia, says his firm could ultimately map the genomes of 50m cancer patients alone a year, helping identify the genetic traits that contribute to the disease. He hopes more data sharing will lead to positive behavioural change in healthcare provision. "My father was sick and for my mother it was a nightmare," he says. "He had repeated MRI scans in different hospitals because they each couldn't access the data. That's a terrible use of healthcare funds. "We need to break down these silos. It's not a tech problem, it's a mentality concern". (*Telegraph*, 16 September 2017)

Besides, H-IoT raises a host of ethical concerns related to the sensitivity of health-related data and to the vulnerability of patients who are unable to control the data flows being generated (Mittelstadt and Floridi 2016; Mittelstadt 2017a, b). Another relevant ethical issue is represented by fairness and accountability in algorithmic decision making. Algorithms self-learn from data that can be biased, which implies, as a consequence, that "algorithms may mirror human biases in decision making" (Char et al. 2018, p. 981) and lead to discriminatory practices (Schadt 2012).

(26) Machine learning algorithms are picking up deeply ingrained race and gender prejudices concealed within the patterns of language use, scientists say. (*Guardian*, 13 April 2017)

In the provision of medical care, introducing algorithms raises questions about the nature of the relationship between physicians and patients and signals a changing paradigm of care, where "Dr Google will see you now" (*Age*, 1 January 2017), further disintermediating doctors. However, "ignoring clinical thinking is dangerous" and algorithms should be viewed "as thinking partners, rather than replacements, for doctors" (Obermeyer and Lee 2017, p. 1211).

(27) Technology firms are already playing a bigger part in health care as phones become more powerful and patients take control of their own diagnosis and treatment. But the more far-sighted hospitals are hoping to remain at the centre of the health-care ecosystem, even as their role changes. (*Economist*, 8 April 2017)

Because it is very expensive, big data research has evolved into a "mixed economy" with both public and private funding (Lipworth et al. 2017, p. 493) but, as seen above in the case of DeepMind, the commercialisation of big data problematises issues of privacy and public interest. Anonymity is also provisional since, as it turns out, de-identified data can be re-identified (example 28) through reverse engineering. In the future, data sharing may increase pressure from private insurance companies.

(28) Richard Dickinson, a partner in data privacy at law firm APKSH [...] warns even anonymised clinical data can be problematic under data protection law if it "can be reverse engineered" to identify the person who supplied it. (*Telegraph*, 16 September 2017)

Health data is considered sensitive data and, as such, its processing should follow high data security and privacy constraints. Instead, the healthcare sector is regularly targeted by "a barrage of cyberattacks that have disputed vital services and exposed vast amounts of sensitive data" (Mohammed 2017, p. 1771). Privacy breaches result in loss of trust. Uses of health data in commercial contexts cause "context collapse" (Marwick and boyd 2010),[9] when people's expectations about the use of their data when dealing with a company are different from their expectations when using public health services, because they are not aware of how their health data might also be used by companies for research. Wearable devices may turn into "dispositifs of capture", favouring surveillant practices "that effectively transform embodied actors/actions into *unbodied* referents, codified flows that subsequently act on the body in various ways" (French and Gavin 2016, p. 20, original emphasis), which are not always in people's best interest.

3.4   CONCLUDING REMARKS

Moving from the investigation of a specialised corpus constructed on the topic of big data in the biomedical field and in healthcare, this chapter has elicited a few recurrent linguistic and discursive strategies currently employed to frame the big data and healthcare debate in the global news media. A number of exploratory probes to retrieve keywords have been followed by the discourse analysis of wider stretches of text and, in particular, a closer focus on the representation of agency by means of periphrastic causative verbs. Two main actors are foregrounded, patients and doctors, with technology and information as helpers that increasingly mediate between them, while healthcare is undergoing a shift towards self-management, from institutional to patient-centred care.

Findings show that the use of big data in healthcare is represented as an opportunity (as regards quality, accessibility, diagnosis, prevention, prediction, public spending savings, fitness and patient-centredness), but also a challenge (as regards privacy, security, collection and storage up to the risk of algorithmic discrimination). It also places an even greater responsibility on patients' shoulders that is somehow forgetful of hard-to-eradicate status and education inequalities.[10]

Nonetheless, the greatest argument in favour of big data is still that it can provide cutting-edge healthcare, with real-time medical data ensuring personalised treatments. Proponents of machine learning point out that healthcare organisations can use electronic medical records and data from wearables to predict diseases, prevent deadly infections[11] and recognise patterns of unknown and unreported adverse side effects of medical products.

Scientific and technological innovation generates both enormous potential benefits and grave risks. It appears, however, that "the collection, storage and analysis of biomedical Big Data potentially raises serious ethical problems, which may threaten the huge opportunities it offers" (Oxford Internet Institute, n.d.). These ethical problems tend to be obfuscated by the multiplicity of interests and investments in big data. A broad social consensus about the uses and the limits of data-driven research in healthcare has still not emerged. When it does, it will have to involve participation from all sectors of society if basic rights are to be adjusted to the big data technology landscape.

Notes

1. The selected news sources are the *Economist*, the *Financial Times*, the *Guardian*, the *Independent*, the *Telegraph*, the *Times* and *Sunday Times* for the United Kingdom; the *Los Angeles Times*, the *New York Times*, *USA Today* and the *Washington Post* for the United States; the *Age*, the *Australian*, the *Australian Financial Review*, the *Canberra Times*, *Guardian Australia* and the *Sydney Morning Herald* for Australia. A few news items from the *HuffPost UK* and *US* and *TimesLIVE* (South Africa) were also added because of their thematic affinity.
2. Topics included wearable technology, cloud computing, videoconferencing, healthcare apps and electronic patient records.
3. As with the previous chapter, the reference corpus was the English Web Corpus 2013 (enTenTen13), preloaded to Sketch Engine, a corpus of general English made up of texts collected from the internet and amounting to over 19 billion words.
4. "Generally, the higher value (100, 1000, ...) of Simple maths focuses on higher-frequency words (more common words), whereas the lower value (1, 0.1, ...) of Simple maths will rather prefer the words with lower frequency (more rare words)" (Sketch Engine, n.d.).
5. ICO is "the UK's independent authority set up to uphold information rights in the public interest, promoting openness by public bodies and data privacy for individuals" (from the ICO website masthead).
6. In 1987, the Picker Institute and Harvard Medical School highlighted eight principles for patient-centred care. They are: Respect for patients' values, preferences and expressed needs; Coordination and integration of care; Information, communication and education; Physical comfort; Emotional support and alleviation of fear and anxiety; Involvement of family and friends; Transition and continuity, and Access to Care.
7. By contrast, lexical causatives are "verbs that encode both the notions of CAUSE and RESULT" (Wolff and Song 2003, p. 285), as for the verb *facilitate* in the following example taken from the corpus: "The company's platforms facilitate more than 70 million patient consultations a year, and it has data stretching back over 20 years" (*Australian Financial Review*, 11 April 2017, n.p.).

8. "Twenty-four-year-old Ian Burkhart suffered a spinal cord injury in a diving accident that left him paralyzed from his shoulders down. Doctors implanted an electrode array in Burkhart's brain in the part of his motor cortex that controls hand movements. Study co-author Chad Bouton used machine-learning algorithms to decode Burkhart's brain activity and use it to stimulate a sleeve of 130 electrodes worn on his forearm" (Lewis 2016, n.p.).

9. In Marwick and boyd's analysis, the concept of "context collapse" originates from the observation that "social media collapse diverse social contexts into one, making it difficult for people to engage in the complex negotiations needed to vary identity presentation, manage impressions, and save face" (2010, p. 123).

10. "Rather than engaging in external understandings of causality in the world, Big Data works on changing social behaviour by enabling greater adaptive reflexivity. If, through Big Data, we could detect and manage our own biorhythms and know the effects of poor eating or a lack of exercise, we could monitor our own health and not need costly medical interventions" (Chandler 2015, p. 843).

11. "The ability of big data to track communicable diseases such as the Zika virus and travel patterns also has huge potential. Big data and analytics have already played a role in containing previous viral outbreaks of cholera and malaria" (*Guardian*, 18 May 2016).

References

Baker, Paul. 2006. *Using Corpora in Discourse Analysis.* London and New York: Continuum.

Baker, Paul, and Tony McEnery, eds. 2015. *Corpora and Discourse Studies: Integrating Discourse and Corpora.* Basingstoke and New York: Palgrave Macmillan.

Baker, Paul, Costas Gabrielatos, Majid Khosravinik, Michał Krzyżanowski, Tony McEnery, and Ruth Wodak. 2008. "A Useful Methodological Synergy? Combining Critical Discourse Analysis and Corpus Linguistics to Examine Discourses of Refugees and Asylum Seekers in the UK Press." *Discourse & Society* 19, no. 3: 273–306. https://doi.org/10.1177/0957926508088962.

Bednarek, Monika, and Helen Caple. 2014. "Why Do News Values Matter? Towards a New Methodological Framework for Analysing News Discourse in Critical Discourse Analysis and Beyond." *Discourse & Society* 25, no. 2: 135–158. https://doi.org/10.1177/0957926513516041.

Burgess, Matt. 2017. "NHS DeepMind Deal Broke Data Protection Law, Regulator Rules." *Wired*, 3 July. http://www.wired.co.uk.

Chandler, David. 2015. "A World Without Causation: Big Data and the Coming of Age of Posthumanism." *Millennium: Journal of International Studies* 43, no. 3: 833–851. https://doi.org/10.1177/0305829815576817.

Char, Danton S., Nigam H. Shah, and David Magnus. 2018. "Implementing Machine Learning in Health Care—Addressing Ethical Challenges." *New England Journal of Medicine* 378, no. 11: 981–983. https://www.nejm.org/doi/full/10.1056/NEJMp1714229.

Chen, Jonathan H., and Steven M. Asch. 2017. "Machine Learning and Prediction in Medicine—Beyond the Peak of Inflated Expectations." *New England Journal of Medicine* 376, no. 26: 2507–2509. https://www.nejm.org/doi/full/10.1056/NEJMp1702071.

Denham, Elizabeth. 2017. "Four Lessons NHS Trusts Can Learn from the Royal Free Case." ICO. Information Commissioner's Blog. 3 July. https://iconewsblog.org.uk.

Duranti, Alessandro. 2004. "Agency in Language." In *A Companion to Linguistic Anthropology*, edited by Alessandro Duranti, 451–473. Malden, MA: Blackwell. https://doi.org/10.1002/9780470996522.ch20.

Fiebig, Denzel G. 2017. "Big Data: Will It Improve Patient-Centered Care?" *The Patient—Patient-Centered Outcomes Research* 10, no. 2: 133–139. https://doi.org/10.1007/s40271-016-0201-0.

Fisher, Eran. 2010. *Media and New Capitalism in the Digital Age: The Spirit of Networks*. New York and Basingstoke: Palgrave Macmillan.

French, Martin, and Gavin J. D. Smith. 2016. "Surveillance and Embodiment: Dispositifs of Capture." *Body & Society* 22, no. 2: 3–27. https://doi.org/10.1177/1357034X16643169.

Garzone, Giuliana, and Francesca Santulli. 2004. "What Can Corpus Linguistics Do for Critical Discourse Analysis?" In *Corpora and Discourse*, edited by Alan Partington, John Morley, and Louann Haarman, 351–368. Bern: Peter Lang.

Gurrin, Cathal, Alan Smeaton, and Aiden R. Doherty. 2014. "LifeLogging: Personal Big Data." *Foundations and Trends® in Information Retrieval* 8, no. 1: 1–125. http://dx.doi.org/10.1561/1500000033.

Haug, Charlotte. 2017. "Whose Data Are They Anyway? Can a Patient Perspective Advance the Data-Sharing Debate?" *New England Journal of Medicine* 376, no. 23: 2203–2205. https://www-nejm-org.pros.lib.unimi.it:2050/doi/pdf/10.1056/NEJMp1704485.

ICO Information Commissioner's Office. 2017. "Royal Free—Google DeepMind Trial Failed to Comply with Data Protection Law." 3 July. https://ico.org.uk.

Iedema, Rick, and Katherine Carroll. 2014. "Intervening in Health Care Communication Using Discourse Analysis." In *Discourse in Context: Contemporary Applied Linguistics*, vol. 3, edited by John Flowerdew, 185–204. London: Bloomsbury.

Kilgarriff, Adam. 2009. "Simple Maths for Keywords." In *Proceedings of the Corpus Linguistics Conference CL2009*, edited by Michaela Mahlberg, Victorina González-Díaz, and Catherine Smith, article number 171, 1–6. Liverpool: University of Liverpool. http://ucrel.lancs.ac.uk/publications/cl2009.

———. 2012. "Getting to Know Your Corpus." In *Text, Speech and Dialogue*. Lecture Notes in Computer Science, vol. 7499, edited by Petr Sojka, Aleš Horák, Ivan Kopeček, and Karel Pala, 3–15. Heidelberg and Berlin: Springer. https://doi.org/10.1007/978-3-642-32790-2_1.

Kilgarriff, Adam, and Gregory Grefenstette. 2003. "Introduction to the Special Issue on the Web as Corpus." *Computational Linguistics* 29, no. 3: 333–347. https://doi.org/10.1162/089120103322711569.

Kilgarriff, Adam, Vít Baisa, Jan Bušta, Miloš Jakubíček, Vojtěch Kovář, Jan Michelfeit, Pavel Rychlý, and Vít Suchomel. 2014. "The Sketch Engine: Ten Years On." *Lexicography* 1, no. 7: 7–36. https://doi.org/10.1007/s40607-014-0009-9.

Koester, Almut. 2010. "Building Small Specialised Corpora." In *The Routledge Handbook of Corpus Linguistics*, edited by Anne O'Keeffe and Michael McCarthy, 66–79. Abingdon and New York: Routledge.

Kreindler, Sara. 2015. "The Politics of Patient-Centred Care." *Health Expectations* 18, no. 5: 1139–1150. https://doi.org/10.1111/hex.12087.

Krucoff, Max, Shervin Rahimpour, Marc W. Slutzsky, V. Reggie Edgerton, and Dennis A. Turner. 2016. "Enhancing Nervous System Recovery through Neurobiologics, Neural Interface Training, and Neurorehabilitation." *Frontiers in Neuroscience* 10: 584. https://doi.org/10.3389/fnins.2016.00584.

Lanzing, Marjolein. 2016. "The Transparent Self." *Ethics and Information Technology* 18, no. 1: 9–16. https://doi.org/10.1007/s10676-016-9396-y.

Lewis, Tanya. 2016. "Neuroprosthesis Restores Arm Movement." *Scientist*, 13 April. https://www.the-scientist.com.

Lipworth, Wendy, Paul H. Mason, Ian Kerridge, and John P. A. Ioannidis. 2017. "Ethics and Epistemology in Big Data Research." *Bioethical Inquiry* 14, no. 4: 489–500. https://doi.org/10.1007/s11673-017-9771-3.

Lomas, Natasha. 2016. "Concerns Raised Over Broad Scope of DeepMind-NHS Health Data-Sharing Deal." *TechCrunch*, 4 May. https://beta.techcrunch.com.

Marr, Bernard. 2015. "How Big Data Is Changing Healthcare." *Forbes*, 21 April. https://www.forbes.com/sites/bernardmarr/2015/04/21/how-big-data-is-changing-healthcare.

Marwick, Alice E., and danah boyd. 2010. "I Tweet Honestly, I Tweet Passionately: Twitter Users, Context Collapse, and the Imagined Audience." *New Media & Society* 13, no. 1: 114–133. https://doi.org/10.1177/1461444810365313.

Mittelstadt, Brent, and Luciano Floridi, eds. 2016. *The Ethics of Biomedical Big Data*. Cham (ZG): Springer.

Mittelstadt, Brent. 2017a. "Ethics of the Health-Related Internet of Things: A Narrative Review." *Ethics and Information Technology* 19, no. 3: 157–175. https://doi.org/10.1007/s10676-017-9426-4.

———. 2017b. "Designing the Health-Related Internet of Things: Ethical Principles and Guidelines." http://dx.doi.org/10.2139/ssrn.2943006.

Mohammed, Derek. 2017. "U.S. Healthcare Industry: Cybersecurity Regulatory and Compliance Issues." *Journal of Research in Business, Economics and Management* 9, no. 5: 1771–1776. http://www.scitecresearch.com/journals/index.php/jrbem/article/view/1347/970.

NHS England. n.d. "Patient Activation." https://www.england.nhs.uk.

Obermeyer, Ziad, and Ezekiel J. Emanuel. 2016. "Predicting the Future—Big Data, Machine Learning, and Clinical Medicine." *New England Journal of Medicine* 375, no. 13: 1216–1219. https://www.nejm.org/doi/full/10.1056/NEJMp1606181.

Obermeyer, Ziad, and Thomas H. Lee. 2017. "Lost in Thought—The Limits of the Human Mind and the Future of Medicine." *New England Journal of Medicine* 377, no. 13: 1209–1211. https://www.nejm.org/doi/full/10.1056/NEJMp1705348.

Oxford Internet Institute. n.d. "Ethics of Biomedical Big Data". https://www.oii.ox.ac.uk.

Partington, Alan. 2004. "Corpora and Discourse, a Most Congruous Beast." In *Corpora and Discourse*, edited by Alan Partington, John Morley, and Louann Haarman, 11–20. Bern: Peter Lang.

Partington, Alan, Alison Duguid, and Charlotte Taylor. 2013. *Patterns and Meanings in Discourse: Theory and Practice in Corpus-Assisted Discourse Studies (CADS)*. Amsterdam and Philadelphia: John Benjamins.

Picker Institute. n.d. "Principles of Patient-Centered Care." http://pickerinstitute.org/about/picker-principles.

Powles, Julia, and Hal Hodson. 2017. "Google DeepMind and Healthcare in an Age of Algorithms." *Health and Technology* 7, no. 4: 351–367. https://doi.org/10.1007/s12553-017-0179-1.

Schadt, Eric E. 2012. "The Changing Privacy Landscape in the Era of Big Data." *Molecular Systems Biology* 8, no. 1, article number 612. https://doi.org/10.1038/msb.2012.47.

Scott, Mike. 1997. "PC Analysis of Key Words—And Key Key Words." *System* 25, no. 2, 233–245. https://doi.org/10.1016/S0346-251X(97)00011-0.

———. 1999. *WordSmith Tools Help Manual* (Version 3.0). Oxford: Mike Scott and Oxford University Press.

Singer, Emily. 2013. "Our Bodies, Our Data." *Quanta Magazine*. 7 October. https://www.quantamagazine.org.

Sittig, Dean F., and Hardeep Singh. 2012. "Rights and Responsibilities of Users of Electronic Health Records." *Canadian Medical Association Journal* 184, no. 13: 1479–1483. https://doi.org/10.1503/cmaj.111599.

Sketch Engine. n.d. *Simple Maths.* https://www.sketchengine.eu/documentation/simple-maths.

Wolff, Phillip, and Grace Song. 2003. "Models of Causation and the Semantics of Causal Verbs." *Cognitive Psychology* 47, no. 3: 276–332. http://dx.doi.org/10.1016/S0010-0285(03)00036-7.

CHAPTER 4

Ethical Issues and Legal Frameworks in the Big Data Debate

Abstract The chapter concentrates on regulatory and legal frameworks in data protection that attempt to respond to ethical issues posed by big data acquisition, processing, storage and use, especially in those instances in which it sidesteps current directives. The main focus is on the European Union's General Data Protection Regulation (GDPR), its inspiring principles, i.e. fairness, accountability and transparency, main actors, i.e. data subjects, controllers and processors, and criticalities. The discourse analysis of the document illustrates the complexity of data protection against the dematerialisation of the economy in the big data ecosystem and the difficulty to harmonise Member States' multiple views on privacy. It explores the discursive overlap of law and ethics and the impact that the legal curb on data processing and repurposing is generating.

Keywords Accountability · Big data · GDPR · Legal discourse · Privacy

4.1 Study Background

The overview of the big data debate has so far touched upon the emergence of new epistemological paradigms and their impact on knowledge production and dissemination, the news media and healthcare, showing the complex ethical issues that big data raises. "Data science provides

M. C. Paganoni, *Framing Big Data*,
https://doi.org/10.1007/978-3-030-16788-2_4

huge opportunities to improve private and public life, as well as our environment (consider the development of smart cities or the problems caused by carbon emissions). Unfortunately, such opportunities are also coupled to significant ethical challenges" (Floridi and Taddeo 2016, p. 1). For this reason, the present chapter is dedicated to the discussion of regulatory and legal data protection frameworks at the supranational level that attempt to respond to the ethical challenges posed by big data collection, processing, storage and use, especially in those instances in which it sidesteps current directives.

To begin with, since big data is "data" in the first place, the related legislation evolves from existing data privacy and data protection laws that have already attempted to address what is ultimately the digital divide and its unequal distribution of knowledge, access and power (van Deursen and Mossberger 2018). At the same time, as several scholars claim, there is an ontological discontinuity between data and big data, due to the variously defined characteristics of the latter that still await assessment (see Kitchin 2014a, b; Kitchin and McArdle 2016). "Given the relatively early point in the present data revolution, it is not at all certain how the present transformations will unfold and settle, and what will be the broader consequences of changes taking place" (Kitchin 2014b, p. xvii). It can be surmised, then, that "the messiness of data [*as*] a reflection of a complexity of nature" (Mazzocchi 2015, p. 1252) is not so easily adapted to conventional legal formats shaped on a matter of a substantially different quality (Zödi 2017). The Council for Big Data, Ethics, and Society expresses the same opinion when it observes that "the proliferation of big data raises ethical issues that demand deliberation. Big data's broad ethical consequences strain the familiar conceptual and infrastructural resources of science and technology ethics" (Metcalf et al. 2016, p. 3).

Since there is no single legislative regime to enforce uniform privacy and data protection policy, the ways in which data and the issues related to its use are formalised in law are consequently different, when not divergent, around the world. "While some countries may construct the private sector as the primary privacy invader, others focus on the policies and actions of public authorities" (Barnard-Wills 2013, p. 171). What is celebrated as Data Protection Day in the European Union on 28 January is called International Privacy Day outside of Europe (eu-LISA 2018).[1]

A further case in point is the current transatlantic privacy debate between the United States and Europe,[2] as to which forms of protections should be accorded to data pertaining to people, since "data privacy has evolved quite differently" (Cobb 2018, n.p.) in the two geopolitical contexts, characterised by different legal cultures and jurisdictions. In the United States, data protection falls under the rubric of privacy laws and, according to a recent op-ed in the *New York Times*, the predominant approach can be summarised as "collect data first, ask questions later" (Burt and Geer 2017, n.p.). In the European Union, data protection legislation is stricter and more extensive for a variety of reasons that are rooted in history and politics (Barnard-Wills 2013; Freude and Freude 2016).[3] It is nonetheless true that Member States have "different historical experiences and contemporary attitudes about data collection", which results in the law "being staggeringly complex" (Cool 2018, n.p.).

Australia is currently calling for a tighter privacy law in the growing awareness that "[the] issue of privacy and its flip side, data security, has moved out of the realm of the IT department to emerge as a geopolitical issue" (Connors 2018, n.p.). Worldwide, the attempt to implement cybersecurity has led some countries (like Brazil) to locally filter and limit access to the web "as a mode of resistance to government agencies' and corporations' increasing encroachments on internet users' privacy", what the *Guardian* renamed as the "Balkanisation of the Internet" (Brown 2013, n.p.).

Until recently, the law governing data protection in the United Kingdom was the Data Protection Act, which was approved in 1998. Since 23 May 2018, however, a post-Brexit Data Protection Act (DPA 2018) has updated privacy laws to respond to the new European Union regulatory framework and address issues that are specific to domestic UK law, such as the processing of data in immigration, for national security, for criminal law enforcement and regarding the powers of the UK Information Commissioner' Office (ICO) itself.

With the awareness that data protection laws are different between countries and legislative regimes (Zödi 2017), the following analysis focalises on the pertinent legislation in the European Union against the background of its pronouncements on data rights, in particular the GDPR, which should be considered the current endpoint of a long development in the legislation.

In 1981, the Council of Europe's Convention 108 inaugurated data protection legislation, rendering the right to privacy a legal imperative at an international level and setting out provisions on transborder data flows. In 1995, the EU adopted the Data Protection Directive

(DPD 95/46/EC), responding to technological development and introducing a new set of definitions such as *processing, sensitive personal data* and *consent*. In 2004, the European Data Protection Supervisor (EDPS) was established, the data protection authority whose task is to ensure that EU institutions, bodies and agencies respect the right to privacy when processing personal data.

The progressive incorporation of human rights into European Union primary law has taken place since the full adoption of the Treaty of Lisbon in 2009 (Clifford and Ausloos 2017). For the first time, under Article 8 of the 2012 Charter of Fundamental Rights of the European Union, data protection has been established as a stand-alone right that includes fair processing, specified purposes, consent or some other legitimate basis, right of access, right to rectification, and control by an independent authority (McDermott 2017, p. 2). On 25 May 2018, the GDPR, a binding legislative act for all EU countries, came into effect repealing the outdated DPD and marking a turning point in the history of data protection.

The GDPR is a regulation[4] whereby the European Parliament, the Council of the European Union and the European Commission intend to strengthen and homogenise data protection for all EU residents, giving them greater control over their data. Much more articulated than the DPD, the GDPR consists of eleven Chapters and ninety-nine Articles, plus one hundred and seventy-three Recitals that add contextual information to the Articles and are essential to practical applications of the Regulation. It is not exclusively aimed at European citizens and de-emphasises notions of nationality and residence in favour of the protection of fundamental rights as stated by EU primary law (Froud 2018). It regulates the processing of personal data, "regardless of whether the processing takes place in the Union or not" (art. 3.1), and wherever "Member State law applies by virtue of public international law" (art. 3.3).

In 1995, the Internet and mobile phones were still at a nascent stage and "Big Data was simply not an issue or at least not in the recent narrative framework" (Zödi 2017, p. 76), whereas the GDPR is steeped in the big data ecosystem. It marks a decided change in European guidelines for its reframing of data protection principles and the power to impose huge administrative fines in case of non-compliance.[5] It allocates monitoring power to independent and public supervisory authorities, "free from external influence, whether direct or indirect" (art. 52.2), which are entrusted by Member States and ultimately report to the Commission, the EDPS and the European Data Protection Board (EDPB).[6] At a more empirical level, because of its effort to strengthen EU citizens' right to data protection, it has significant consequences for the corporate world globally.[7] On the very day it came into effect, not unsurprisingly, Google, Facebook, WhatsApp and Instagram were sued for obtaining forced consent (Meyer 2018, n.p.).

For the purpose of this investigation of big data, the GDPR is of help in understanding the issues lawmakers and policymakers intend to address by modifying and updating previous guidelines and, conversely, to get a sense of those grey areas that are not yet adequately mapped, with special attention to ethical issues. It has to be stated again that big data innovation is not exempt from an intrinsic ambivalence, in so far as technology liberates and coerces people simultaneously, so that "they gain personal benefit at the same time as they become enmeshed in a system that seeks to gain from their participation" (Kitchin 2014b, p. 165).

A major ethical stumbling block to the use of big data is that it allegedly discards traditional notions of moral agency and free will "by reducing knowable outcomes of actions, while increasing unintended consequences" (Zwitter 2014, p. 3). Connected to this issue is the widely debated question of whether intelligent decision-making algorithms produce unbiased results across big data contexts (Mittelstadt et al. 2016). Moreover, informational privacy should always be derived from human dignity (Floridi 2016), while the notion of group privacy is progressively taking shape (Floridi 2017). In a nutshell, a large spectrum of concepts, behaviours and practices still await redefined ethics codes in the areas of data generation, collection, mining and analysis (Metcalf and Crawford 2016; Veale and Binns 2017).

4.2 Materials and Methods

The following analysis is primarily focused on the GDPR. The text of the Regulation was published in the *Official Journal of the European Union* on 4 May 2016, and became binding on 25 May 2018, after a two-year transition period. The analysis intends to identify the linguistic devices and discursive strategies by means of which it addresses data protection and to reflect on how its salient points attempt to provide a legal framework of emerging ethical issues that are peculiar to, or have been intensified by, the new big data ecosystem (Taylor et al. 2017).

The selected methodology for this kind of inquiry espouses Discourse and Critical Discourse Analysis (Tannen et al. 2015; Flowerdew and Richardson 2018). The aim is to carry out a linguistically informed inspection at the textual microlevel and to connect the inferred meanings of lexical, syntactic and discursive strategies with the macrolevel of social practice in order to capture the echo of societal debates in the encoding of tacit and explicit ideological meanings.

In this specific case, the discourse analysis of the Regulation is set against the background of the geopolitical context, specific legal frameworks, i.e. those of data protection in EU agenda-setting, and power (im)balance in the European Union. Critical insights into data protection and its discursive formulations and reformulations have been drawn from EU legal and policy literature (Craig 2015; Clifford and Ausloos 2017; Selbst and Powles 2017; Forgó et al. 2017; Wachter et al. 2017) and from the ICO's reports and updates (ICO 2017, 2018a, b).

4.3   A Discourse-Analytic Focus on the GDPR

The scope of the GDPR covers "the protection of natural persons with regard to the processing of personal data and rules relating to the free movement of personal data" (art. 1.1), seen as one of the "fundamental rights and freedoms of natural persons" (art. 1.2). From the start, it can be observed how emphasis is placed on personal data in connection with human rights, a multi-layered concept conventionally summarised in the binomial *rights and freedoms*, occurring thirty-eight times in the document, with nine instances in which it collocates with *(high) risk*. At least in principle, the protection of fundamental rights and freedoms does not clash with the free circulation of data in the European Union as long as data is kept secure and private.

Although it does not contain any direct mention of big data, but just of data, the GDPR expands previous data protection rules to address big data issues. The encoding of specific lexical items provides a clue to this new context. References to "sets of personal data" (art. 4.2), the "processing on a large scale of special categories of data" (art. 27.2.a, 35.3.b, 37.1.c) and the use of "new technologies" (art. 35.1), together with the twenty-four occurrences of *personal data breach(es)*, clearly allude to the informational volume of big data and the ominous implications of its mishandling. A case in point is the combination of distinct sets of highly sensitive data, which can potentially lead to predictive privacy harms or data discrimination. This awareness explains the presence of terms like *profiling* (22 occurrences), *pseudonymisation*, i.e. deidentification (6), and *encryption* (3) that are commonly used to describe typical data and IT security technologies. Though already used in the 1995 DPD, *data flows* and *unauthorised disclosure* are now charged with implications from the post-Cambridge Analytica world.

The GDPR mentions genetic, biometric and health data (art. 4.1-13-14-15), i.e. any data relating to inherited or acquired genetic characteristics or disclosing unique information about a natural person's identity, physiology or health that allows the unique identification of that person to be confirmed. Further in the text, special categories of data are defined as "personal data revealing racial or ethnic origin, political opinions, religious or philosophical beliefs, or trade union membership, [...] genetic data, biometric data [...], data concerning health or data concerning a natural person's sex life or sexual orientation" (art. 9.1). In Recital 10, these special categories of data are defined as "sensitive data". It follows that, since they are "particularly sensitive in relation to fundamental rights and freedoms", special categories of data are subject to greater restrictions, as "the context of their processing could create significant risks to the fundamental rights and freedoms" (Recital 51).

From the start, the comparison of key definitions between the DPD and the GDPR illustrates the much more articulated nature of the latter. Of the eight definitions (*personal data* and *data subject, data processing, filing system, controller, processor, third party, recipient, the data subject's consent*) provided in Article 2 of the DPD and carried over to the GDPR, only *filing system, processor* and *third party* have stayed the same. The six remaining definitions have been updated to adjust to the EU's evolution and the big data ecosystem (Table 4.1), and eighteen new ones added.

The GDPR protects personal data of "an identified or identifiable natural person", called the "data subject" (art. 4.1), with special attention in the case of children. Data can be processed wholly or partly by automated means besides manual filing systems (art. 2.1). Personal data includes "location data and online identifiers" (art. 4.1), provided by "devices, applications, tools and protocols, such as internet protocol addresses, cookie identifiers or other identifiers such as radio frequency identification tags" (Recital 30). This provision states that digital footprints (i.e. location data, browsing history, access to social media etc.) should be recognised as part of the data subject's personal information.

Table 4.1 Summary of key definitions in the DPD and in the GDPR

1995 Data Protection Directive (DPD)	*2018 General Data Protection Directive (GDPR)*
Personal data: "any information relating to an identified or identifiable natural person ('data subject'); an identifiable person is one who can be identified, directly or indirectly, in particular by reference to an identification number or to one or more factors specific to his physical, physiological, mental, economic, cultural or social identity" (Ch. 1, art. 2.a)	*Personal data*: "any information relating to an identified or identifiable natural person ('data subject'); an identifiable natural person is one who can be identified, directly or indirectly, in particular by reference to an identifier such as a name, an identification number, <u>location data, an online identifier</u> or to one or more factors specific to the physical, physiological, <u>genetic</u>, mental, economic, cultural or social identity of that natural person" (art. 4.1)
Data processing: "any operation or set of operations which is performed upon personal data, whether or not by automatic means, such as collection, recording, organization, storage, adaptation or alteration, retrieval, consultation, use, disclosure by transmission, dissemination or otherwise making available, alignment or combination, <u>blocking</u>, erasure or destruction" (Ch. 1, art. 2.b)	*Data processing*: "any operation or set of operations which is performed on personal data or <u>on sets of personal data</u>, whether or not by automated means, such as collection, recording, organisation, <u>structuring</u>, storage, adaptation or alteration, retrieval, consultation, use, disclosure by transmission, dissemination or otherwise making available, alignment or combination, <u>restriction</u>, erasure or destruction" (art. 4.2)
Data controller: "the natural or legal person, public authority, agency or any other body which alone or jointly with others determines the purposes and means of the processing of personal data; where the purposes and means of processing are determined <u>by national or Community laws</u> or regulations, the controller or the specific criteria for his nomination may be designated <u>by national or Community law</u>" (Ch. 1, art. 2.d)	*Data controller*: "the natural or legal person, public authority, agency or other body which, alone or jointly with others, determines the purposes and means of the processing of personal data; where the purposes and means of such processing are determined <u>by Union or Member State law</u>, the controller or the specific criteria for its nomination may be provided for <u>by Union or Member State law</u>" (art. 4.7)
Recipient: "a natural or legal person, public authority, agency or any other body to whom data are disclosed, whether a third party or not; however, authorities which may receive data in the framework of a particular inquiry shall not be regarded as recipients" (Ch. 1, art. 2.g)	*Recipient*: "a natural or legal person, public authority, agency or another body, to which the personal data are disclosed, whether a third party or not. However, public authorities which may receive personal data in the framework of a particular inquiry <u>in accordance with Union or Member State law</u> shall not be regarded as recipients; the processing of those data by those public authorities shall be in compliance with the applicable data protection <u>rules according to the purposes of the processing</u>" (art. 4.9)

(continued)

Table 4.1   (continued)

1995 Data Protection Directive (DPD)	2018 General Data Protection Directive (GDPR)
Consent: "any freely given specific and informed indication of his wishes by which the data subject signifies his agreement to personal data relating to him being processed" (Ch. 1, art. 2.h)	*Consent*: "any freely given, specific, informed and <u>unambiguous</u> indication of the data subject's wishes by which he or she, <u>by a statement or by a clear affirmative action</u>, signifies agreement to the processing of personal data relating to him or her" (art. 4.11)

The "controller" is defined as "the natural or legal person, public authority, agency or other body which, alone or jointly with others, determines the purposes and means of the processing of personal data" (art. 4.7). The "processor" is "a natural or legal person, public authority, agency or other body which processes personal data on behalf of the controller" (art. 4.8), as when a company outsources to a marketing agency. The GDPR applies to the European Economic Area (EEA), which includes Iceland, Liechtenstein and Norway besides the twenty-eight Member States. When personal data is transferred outside the EEA, the safeguards offered under the Regulation are maintained. In sum, with a territorial scope that manages to include extraterritorial organisations, the GDPR protects the processing of personal data, "regardless of whether the processing takes place in the Union or not" (art. 3.1), and wherever "Member State law applies by virtue of public international law" (art. 3.3). It does not apply in cases of "national security", "defence", "public security", "criminal offences", and "important objectives of general public interest of the Union or of a Member State" (art. 23). Derogations from the prohibition on processing special categories of personal data may be made "for reasons of substantial public interest" (art. 9.2.g), when "it is in the public interest to do so" (Recital 52), i.e. in the area of public health.

4.3.1   The GDPR Inspiring Principles

Although privacy is never mentioned once in the GDPR, data protection stems from the right to privacy, one of the fundamental rights that are safeguarded by the Regulation and that was established as a standalone right under Article 8 of the 2012 Charter of Fundamental Rights of the European Union (McDermott 2017). For this reason, all the provisions that are enshrined in the GDPR aim to protect personal data "by default".[8] The processing of personal data follows seven inspiring principles: "lawfulness, fairness and transparency; purpose limitation; data minimisation; accuracy; storage limitation; integrity and confidentiality" (art. 5.1), and "accountability" (art. 5.2).

That "any processing of personal data must be lawful and fair to the individuals concerned" (art. 38) was a provision already set out in the DPD. The GDPR introduces the data processor's direct liability for intentional or negligent infringement of provisions (art. 83.3), blames personal data breaches for leading to "accidental or unlawful destruction, loss, alteration, unauthorised disclosure of, or access to, personal data transmitted, stored or otherwise processed" (art. 4.12, 32.2), and stresses the data subject's right to judicial redress and administrative compensation for "the damage suffered" (art. 82.1 and 5, 83.2.a and c). The principle of fairness protects data subjects from being disenfranchised, or even deceived, by data misuse. In sum, "enforceable and effective rights" (Recitals 108 and 114) for data subjects correspond to "binding and enforceable commitments" (art. 40.3, 42.2, 46.2.e and f) for controllers and processors.

Lawfulness means that "you must have a valid lawful basis in order to process personal data" (ICO 2018b, p. 10). However, in their extensive systematic reading of the GDPR within the EU regulatory framework on data protection, Clifford and Ausloos (2017) argue that the ways in which *fairness* and *transparency* are discursively positioned with regard to lawfulness are shrouded in vagueness. Their claim is that it "is unclear whether fairness was intended as one principle [*or*] two principles (i.e. lawfulness and transparency to be viewed through the lens of an overarching notion of fairness)" (Clifford and Ausloos 2017, pp. 8–9). Other linguistic and discursive cues in the Regulation, however, would seem to legitimate the interpretation of fairness as a separate principle, endowed with the power of "protection of data subjects from controller abuse" (Clifford and Ausloos 2017, p. 12), especially as concerns "fair and transparent processing" (art. 13.2, 14.2, 40.2.a) of personal data by way of algorithms (example 1).

(1) In order to ensure fair and transparent processing in respect of the data subject, taking into account the specific circumstances and context in which the personal data are processed, the controller should use appropriate mathematical or statistical procedures for the profiling, implement technical and organisational measures appropriate to

ensure, in particular, that factors which result in inaccuracies in personal data are corrected and the risk of errors is minimised, secure personal data in a manner that takes account of the potential risks involved for the interests and rights of the data subject and that prevents, inter alia, discriminatory effects on natural persons on the basis of racial or ethnic origin, political opinion, religion or beliefs, trade union membership, genetic or health status or sexual orientation, or processing that results in measures having such an effect. Automated decision-making and profiling based on special categories of personal data should be allowed only under specific conditions. (Recital 71)

Recital 71 refers in particular to fairness-aware data mining technologies and organisational procedures (Veale and Binns 2017, p. 3). As for transparency (art. 5.1.a), which was mentioned only once in the DPD, this is the first time that this ethical principle has been expressly included in a EU regulation (Clifford and Ausloos 2017, p. 23). Transparent data processing is a way "to safeguard the data subject's human dignity" (art. 88.2). Transparency reverberates in the choice of using "clear and plain language" (three occurrences in the text), especially when consent is provided (art. 7.2, example 2), when information is conveyed, to children in particular (art. 12.1),[9] and when communicating a personal data breach to the data subject "without undue delay" (art. 34. 1).

(2) If the data subject's consent is given in the context of a written declaration which also concerns other matters, the request for consent shall be presented in a manner which is clearly distinguishable from the other matters, in an intelligible and easily accessible form, using clear and plain language. (art. 7.2)

A caveat, nonetheless: data protection is context-dependent and data transparency has political importance that may be made instrumental to not-so-transparent dealings. Whenever the ethos of transparency becomes an ideology of indiscriminate openness in science and governance, it can be "weaponised" for partisan reasons that are detrimental to the advancement of knowledge and good governance and favour "regulated industries, lobbyists, and trade organizations" (Levy and Johns 2016, p. 2).[10] By contrast, "transparency can also conflict with efforts to protect privacy. And in some cases, disclosing too much information about how an algorithm works might allow people to game the system" (Courtland 2018, p. 360).

Purpose limitation (art. 5.2.b) states that data should not be further processed in a manner that is incompatible with the purposes for which it has been originally collected. Consequently, organisations or public bodies that intend to use personal data gathered in the European Union must first specify what that data is going to be used for. The provision is probably a step towards the long-awaited regulation of digital giant companies on global markets, including Europe, with data controllers bearing the major burden of responsibility. However, this principle would seem to run counter to one of the main value generating qualities of big data, i.e. that they can be repurposed for different analytical aims, and to possibly hamper advancements in the scientific and medical fields.

In fact, the concept of purpose—the lemma occurs one hundred and twenty-five times in the Regulation—remains problematic. First of all, purpose may be distorted by "context collapse" (Marwick and boyd 2011), i.e. the data subject's multiple and somewhat indiscriminate allocation of trust between the personal and the professional, the public sector and the corporate world, arising from lack of informed consent. Although scientific research "should also include studies conducted in the public interest in the area of public health" (Recital 159), "[i]t is often not possible to fully identify the purpose of personal data processing for scientific research purposes at the time of data collection" (Recital 33). Data subjects are allowed "to give their consent to certain areas of scientific research when in keeping with recognised ethical standards for scientific research" (Recital 33). In a nutshell, "privacy, which the data protection regulations including the purpose limitation principle seek to realize, may not only be seen as a hindering factor for economy and science" and should be highlighted, instead, as "a core value and a necessity of Big Data" (Forgó et al. 2017, p. 40).

Along these lines, *data minimisation* (art. 5.2.c) refers to the use of data, which should be limited to what is necessary in relation to the purposes for which it has been processed. Inaccurate personal data should be erased or rectified without delay. *Accuracy* (art. 5.2.d) requires controllers to ensure that data is kept correct and allows data subjects to update their data when required. *Storage limitation* (art. 5.2.e) enforces data anonymisation when data is no longer necessary for the purposes for which it has been processed. *Integrity and confidentiality* (art. 5.1.f) are summarised in the notion of "appropriate security", which means processing data in a manner that will prevent misuse, hacking and illegal manipulation.

In the GDPR, in fact, the foregrounding of these principles should be set against the several grey areas in the use of big data that are subsumed under the category of *risk*.[11] Ethical values are indeed perceivable in watermark in the protection of "the rights and freedoms of natural persons", but viewed from the more empirical lens of risk assessment and redressive actions (examples 3 and 4), such as mandatory data breach notification within seventy-two hours of first having become aware of the breach, unless the data is strongly encrypted (art. 33.1).

> (3) <u>Taking into account</u> the state of the art, the costs of implementation and the nature, scope, context and purposes of processing as well as <u>the risk of varying likelihood and severity for the rights and freedoms of natural persons</u>, the controller and the processor shall implement appropriate technical and organisational measures to ensure a level of security appropriate to the risk. (art. 33.1)

> (4) When <u>the personal data breach is likely to result in a high risk to the rights and freedoms of natural persons</u>, the controller shall communicate the personal data breach to the data subject without undue delay. (art. 34.1)

Nominalisations (like *risk* and *data breach* above and, elsewhere, *alteration, damage, disclosure, infringement* and *loss*) suit the register of legal discourse and the aims of generalisation and objectivity that it pursues. Inevitably, however, these discursive strategies obfuscate the responsibility and intentionality of noncompliant actors that may instrumentally tweak regulations, especially in what is still a largely underregulated area. In the GDPR, agency (indexed by the use of the deontic *shall* in active sentences) appears to be mostly distributed among ruling bodies (the *Commission*, the *European Data Protection Board, supervisory authorities* and *Member States*) and then granted to controllers and processors. Data subjects, instead, are more frequently the object or the complement of prepositional phrases (examples 6–7).

> (5) <u>The Commission and supervisory authorities</u> shall take appropriate steps to develop international cooperation mechanisms to facilitate the effective enforcement of legislation for the protection of personal data. (art. 50.a)

(6) <u>The controller</u> shall, at the time when personal data are obtained, provide <u>the data subject</u> with all of the following information. (art. 13.1)

(7) <u>Each supervisory authority</u> shall on its territory: upon request, provide information <u>to any data subject</u> concerning the exercise of their rights. (art. 57.1.e)

(8) <u>Member States</u> shall lay down the rules on other penalties applicable to infringements of this Regulation. (art. 84.1)

In conclusion, participants have neatly assigned roles and hierarchical responsibilities in a data landscape that resembles more an "oligopticon" (Kitchin 2014b, p. 133) than a panopticon. Against this background, the mishandling of data is contingent upon the "messy" ontologies of big data itself and the new phenomena it generates: datafication, dataveillance,[12] data footprints and shadows,[13] privacy and data security breaches, profiling, predictive analytics and anticipatory governance, practices of exclusion like social sorting and redlining, control creep,[14] technocratic and corporate governance, as well as issues of digital ownership and intellectual property rights (Kitchin 2014b, *passim*).

The last principle, i.e. *accountability* (art 5.2), is central to the debate about data protection that is currently taking place at the level of EU secondary legislation and will be briefly discussed in what follows. The textual organisation of the Regulation, which separates it from the other principles, would seem to show the elevated status and reach of accountability itself. In actual fact, the accountability principle, which is ubiquitous in contemporary discussions on governance policy—in the European Union as well in the United States (Bass 2015; Craig 2015)— is introduced as a corrective measure to curb the power imbalance between controllers and processors and describes the specific rules and obligations to which both are subject. In this specific case (examples 9 and 10), the GDPR requires data controllers and processors to:

(9) <u>be responsible for</u>, and <u>be able to demonstrate compliance</u> with [principles relating to processing of the personal data]. (art. 5.2)

(10) implement appropriate technical and organisational measures to ensure and to be able to demonstrate that processing is performed in accordance with this Regulation. (art. 24.1)

To be noted, accountability involves not only compliance with the ethical principles of "lawfulness, fairness and transparency", but also an engrained obligation to demonstrate such compliance, embodying ethical behaviour "in *performing* according to the GDPR [...] and in *explaining* this performance to others" (Hijmans and Raab 2018, p. 11, original emphasis), even when not explicitly required by the GDPR. It amounts to saying that corporate responsibility should be designed as an intrinsic ethical stance in business operations and, for example, implemented by enabling technology that will protect privacy from the start (data protection "by design") and not retrospectively.[15] Arguably, however, the ability to demonstrate compliance in the procedural fairness assessment is still discursively framed so as to leave the "burden of proof incumbent upon data subjects" (Clifford and Ausloos 2017, p. 29). Moreover, "[t]he scope of GDPR provisions that might give the public insight into algorithms and the ability to appeal is also in question" (Courtland 2018, p. 360, quoting Brent Mittelstadt, a data ethicist at the Oxford Internet Institute). In sum, if not properly addressed by policymakers, this power imbalance will make of fairness (and accountability) "a poisoned apple" (Clifford and Ausloos 2017, p. 29).

"Facebook users did not consent to have their profiles and online activities acquired, sold, and used to influence activism and votes on behalf of a political campaign" (Salmons 2017b, n.p.). Ending the days of pre-selected opt-in boxes and responding to an increasing need in society, the GDPR imposes stricter rules around consent. The key points of the conditions for consent in Article 7 are also imbued with a sense of ethics and moral concerns. Consent must be freely given, specific, informed and unambiguous about the data subject's wishes.

Besides, the right includes the withdrawal of consent "at any time" (art. 7.3). It follows that companies will have to devise options to activate and withdraw consent at all times, making language clear, accessible and age-appropriate (example 11).

(11) The request for consent shall be presented in a manner which is clearly distinguishable from the other matters, in an intelligible and easily accessible form, using clear and plain language. (art. 7.2)

Since explicit informed consent cannot be assumed but should be the result of "affirmative action" (art. 4.11), clearly separated from other terms and conditions, the provision rules out confusing and complicated terms of service or End User Licensing Agreements.[16]

In the light of the recent episodes of predictive profiling and political manipulation (Puaschunder 2017, 2018), it is to be hoped that informed consent will help protect democracy. The same feeling imbues other societal sectors that are invoking informed consent such as academia, in which big data epistemologies are rewriting the code of ethics and scholars question the ethical implications for online data collection (Schadt 2012; Salmons 2017a, b). Eschewing context collapse in their analytical practices (Marwick and boyd 2011; Davis and Jurgenson 2014), researchers should "protect human subjects and their multiple identities and representations in cyberspace" (Salmons 2017a, p. 128).

4.3.2   The Data Subject's Rights

As in the case of informed consent, the most interesting aspects of the rights enshrined in the GDPR are consistent with the empowerment of data subjects. In particular the right of information about the processing of personal data (art. 13 and 14), the right of access (art. 15), the right to rectification (art. 16), the right to erasure (art. 17), the right to restrict processing (art. 18), the right to data portability (art. 20), the right to object (art. 21), and the right not to be subject to a decision based solely on automated processing, including profiling (art. 22).

(12) The controller shall provide the data subject with the following <u>information necessary to ensure fair and transparent processing in respect of the data subject</u>. (art. 13.1)

(13) <u>The data subject shall have the right to obtain</u> from the controller confirmation as to whether or not personal data concerning him or her are being processed, and, where that is the case, <u>access to the personal data</u> and the following information. (art. 15.1)

(14) <u>The data subject shall have the right to obtain from the controller without undue delay the rectification of inaccurate personal data concerning him or her</u>. Taking into account the purposes of the processing, the data subject shall have the right to have incomplete personal data completed, including by means of providing a supplementary statement. (art. 16)

Characteristic of the information age, the right to erasure (art. 17), also known as right to be forgotten, entitles data subjects to have controllers erase their personal data, stop further dissemination of their data and potentially block third parties from processing their data. The conditions for erasure include that the data is no longer relevant to the original purposes for processing, or that data subjects withdraw consent. It should also be noted that this right requires controllers to compare the subjects' rights to the public interest in the availability of the data when considering the request. "Reasons of public interest in the area of public health" are contemplated (art. 17.3.b).

The right to restrict processing grants data subjects the power to reclaim control over their data, provided that the legitimate grounds of the controller do not override those of the data subject (art. 18.1.d). Data portability allows data subjects to reclaim their personal data, which they have previously made available in a "commonly used and machine readable format" (art. 20.1), and to transmit it to another controller. Finally, the right to object grants data subjects the power to oppose to the processing of their personal data, unless there are "compelling legitimate grounds for the processing which override the interests, rights and freedoms of the data subject or for the establishment, exercise or defence of legal claims" (art. 21.1).

Article 22, which mandates "a right of explanation of all decisions made by automated or artificially intelligent algorithmic systems" (Wachter et al. 2017, p. 1), is currently quite controversial.

(15) The data subject shall have the right not to be subject to a decision based solely on automated processing, including profiling, which produces legal effects concerning him or her or similarly significantly affects him or her. (art. 22.1)

According to a group of scholars of the Oxford Internet Institute with a research track in big data and data ethics, this right does not exist at the moment. The vague language that sustains the provision hides conceptual and procedural gaps that ignore the latest technological advances like artificial intelligence, deep learning, bots and fake news. The point they make is that "even if a right to explanation is legally granted in the future, the feasibility and practical requirements to offer explanations to data subjects remain unclear" (Wachter et al. 2017, p. 46). Contrariwise, critically responding to their analysis, Selbst and Powles (2017) claim the right to explanation in Article 22 is not illusory, "should be interpreted functionally, flexibly, and should, at a minimum, enable a data subject to exercise his or her rights under the GDPR and human rights law" (p. 233).

4.4   Concluding Remarks

The discourse analysis of the GDPR has mainly focused on how the choice of lexical items and the deployment of syntactic strategies in the representation of agency may shed light on a number of innovative aspects in the Regulation. It has outlined how the GDPR principles are discursively framed as "enforceable rights" (art. 47.1.b) that originate from ethical concerns "that were not previously so prominent" (Hijmans and Raab 2018, p. 2). The overlap of law and ethics has been explored to illustrate some of the complexities of data protection against the background of the knowledge society and the dematerialisation of the economy. In the big data ecosystem, the aim of the GDPR is to ensure the same level of data protection in the European Union and "to prevent the EU becoming an unethical data 'haven'" (Madge 2018, n.p.), while still achieving a "thriving data-driven economy" (Zödi 2017, p. 80).

In its attempt to respond to the discontinuities and unique challenges of the big data ecosystem, however, the Regulation appears general and complicated at the same time. With regard to this, the contested views on Article 22 in the GDPR well exemplify how the entire Regulation "can be a toothless or powerful mechanism to protect data subjects depending on its eventual legal interpretation" (Wachter et al. 2017, p. 46). In other words, future directions will have to be devised in the political arena. The extent to which data protection policies will seriously address the revolution brought about by big data will eventually be decided by ruling bodies—the Commission, the EDPB and national supervisory authorities. As the previous analysis has highlighted, all these institutions are discursively prioritised throughout the document.

To conclude, a regulation like the GDPR that needs to harmonise the multiple views of EU Member States is probably unfit to tackle the "messiness" of big data in a comprehensive way and with the right amount of knowledge to fully understand innovation, as the contrasting views of the right to explanation reveal. Nonetheless, the GDPR is a remarkable indicator of the acceleration generated by big data and of the kind of responses it stimulates from different fields of society.

NOTES

1. On 28 January 1981, "the Convention 108 – Convention for the Protection of Individuals with regard to Automatic Processing of Personal Data – was signed. [*It*] is the first binding international instrument to protect an individual against abuses of the processing of their personal information. Moreover, it also imposes some restrictions on transborder flows of personal data to countries where legal regulation does not provide equivalent protection" (eu-LISA 2018, n.p.). On 18 May 2018, the Council of Europe passed an Amending Protocol to update it. The modernised Convention is known as Convention 108⁺.

2. In a column for the *Los Angeles Times* on 5 January 2018, David Lazarus argues that US citizens and consumers are underprotected in comparison with European citizens, one of the reasons being that "Republican lawmakers have blocked all legislation aimed at improving privacy protection or holding companies more accountable for the loss of people's info" (Lazarus 2018, n.p.).

3. For example, among the reasons of "the distrust and revulsion Germans feel toward state surveillance, which help explain the widespread belief that privacy merits special protection", there are memories of the Nazi regime and of the Stasi, the East German secret police (Freude and Freude 2016, p. 2). In fact, in EU regulations, "all personal data is protected by default", while "the privacy interests of individuals in the US are protected in a wide range of situations [*but*] the protection is far from universal" (Cobb 2016, p. 6).

4. According to EU law, "a 'regulation' is a binding legislative act. It must be applied in its entirety across the EU". [...] A 'directive' is a legislative act that sets out a goal that all EU countries must achieve" (European Union, n.d.).

5. In the case of non-compliance with key provisions of the GDPR, regulators have the authority to levy a fine on data controllers and processors. There are two levels of fines according to the nature of infringement. The lower level is "up to €10 million, or 2% of the worldwide annual revenue of the prior financial year, whichever is higher", the upper level is "up to €20 million, or 4% of the worldwide annual revenue of the prior financial year, whichever is higher" (GDPR EU.org, n.d.). "Examples that fall under this [*latter*] category are non-adherence to the core principles of processing personal data, infringement of the rights of data subjects and the transfer of personal data to third countries or international organizations that do not ensure an adequate level of data protection" (*GDPR.Report*, 16 June 2017).

6. The EDPB has been established to ensure a consistent application of the GDPR in the Union.

7. In fact, the application of the GDPR delimits the influence of giant platforms, from social media to high tech and e-commerce that may incur in penalties and fines in case of non-compliance. Besides the US-based companies of Facebook, Apple, Microsoft, Google and Amazon, China's corporate giants like Alibaba, Huawei and Tencent are also affected.
8. "In the EU, data that pertains to you as an identifiable individual is protected, by default, from inception" (Cobb 2018, n.p.).
9. Wang (2017) recounts of a lawyer that rewrote Instagram's terms of use in plain English to make them accessible to children.
10. Levy and Johns (2016) discuss two instances in the US recent history in which, "framed as an unalloyed good (provided that privacy interests can be adequately protected), in practice [*data transparency*] provides a means through which diverse stakeholders attempt to achieve diverse political goals" (p. 2).
11. "The GDPR is a risk-based regulatory framework which relies heavily on controller 'responsibilisation' in the form of data protection by design and by default" (Clifford and Ausloos 2017, p. 9).
12. "Whereas surveillance presumes monitoring for specific purposes, dataveillance entails the continuous tracking of (meta)data for unstated preset purposes" (van Dijck 2014, p. 205).
13. Data footprints are an individual's set of digital activities, while a data shadow describes information about a person which is generated by others (Kitchin 2014b).
14. "Control creep is when the data generated for a form of governance are used for another" (Kitchin 2014b, p. 178).
15. Privacy by design calls for the inclusion of data protection from the onset of the designing of systems rather than a subsequent addition.
16. "Buried within the consents online users click through without ever reading is their explicit approval to allow companies to leverage for whatever purpose they deem appropriate any and all personal information" (Schadt 2012, p. 2). The GDPR states instead that "[u]tmost account shall be taken of whether, inter alia, the performance of a contract, including the provision of a service, is conditional on consent to the processing of personal data that is not necessary for the performance of that contract (art. 7.4)".

References

Barnard-Wills, David. 2013. "Security, Privacy and Surveillance in European Policy Documents." *International Data Privacy Law* 3, no. 3: 170–180. https://doi.org/10.1093/idpl/ipt014.

Bass, Gary D. 2015. "Big Data and Government Accountability: An Agenda for the Future." *I/S: A Journal of Law and Policy* 11, no. 1: 13–48. https://kb.osu.edu/dspace/handle/1811/75432.

Brown, Ian. 2013. "Will NSA Revelations Lead to the Balkanisation of the Internet?" *The Guardian*, 1 November. https://www.theguardian.com/world/2013/nov/01/nsa-revelations-balkanisation-internet.

Burt, Andrew, and Dan Geer. 2017. "The End of Privacy." *New York Times*, 5 October. https://www.nytimes.com/2017/10/05/opinion/privacy-rights-security-breaches.html.

Charter of Fundamental Rights of the European Union. 2012. *Official Journal of the European Union* C 326 (26 October): 391–407. https://eur-lex.europa.eu/legal-content/EN/TXT/?uri=celex:12012P/TXT.

Clifford, Damian, and Jef Ausloos. 2017. "Data Protection and the Role of Fairness." CiTiP Working Paper 29/2017. Centre for IT & IP Law: KU Leuven. https://ssrn.com/abstract=3013139.

Cobb, Stephen. 2016. "Data Privacy and Data Protection: US Law and Legislation." ESET White Paper. ESET. https://www.welivesecurity.com/wp-content/uploads/2018/01/US-data-privacy-legislation-white-paper.pdf.

———. 2018. "Data Privacy vs. Data Protection: Reflecting on Privacy Day and GDPR." *WeLiveSecurity*, 25 January. https://www.welivesecurity.com/2018/01/25/data-privacy-vs-data-protection-gdpr.

Connors, Emma. 2018. "Microsoft President Brad Smith Wants to Save the World from Cyber Warfare." *The Australian Financial Review* (31 May). https://www.afr.com/afr-special/microsoft-president-brad-smith-on-the-dangers-of-a-connected-world-20180508-h0zs3c.

Cool, Alison. 2018. "Europe's Data Protection Law Is a Big, Confusing Mess." *New York Times*, 15 May. https://www.nytimes.com/2018/05/15/opinion/gdpr-europe-data-protection.html.

Council of Europe. 1981. "Convention for the Protection of Individuals with regard to Automatic Processing of Personal Data." European Treaty Series no. 108 (28 January). https://www.coe.int/en/web/conventions/full-list/-/conventions/treaty/108.

Courtland, Rachel. 2018. "The Bias Detectives." *Nature* 558, no. 7710 (20 June): 357–360. https://www.nature.com/articles/d41586-018-05469-3.

Craig, Paul. 2015. "Accountability." In *The Oxford Handbook of European Union Law*, edited by Anthony Arnull and Damian Chalmers, 431–454. Oxford: Oxford University Press.

Data Protection Directive. 1995. "Directive 95/46/EC." *Official Journal of the European Communities* 281 (23 November): 31–50.

Davis, Jenny L., and Nathan Jurgenson. 2014. "Context Collapse: Theorizing Context Collusions and Collisions." *Information, Communication & Society* 17, no. 4: 476–485. https://doi.org/10.1080/1369118X.2014.888458.

eu-LISA (European Agency for the Operational Management of Large-Scale IT Systems in the Area of Freedom, Security and Justice). 2018. "Data Protection Day 2018." 29 January. https://www.eulisa.europa.eu/Newsroom/News/Pages/Data-Protection-Day-2018.aspx.

European Union. n.d. "Regulations, Directives and Other Acts." https://europa.eu/european-union/eulaw/legal-acts_en.

Floridi, Luciano. 2016. "On Human Dignity as a Foundation for the Right to Privacy." *Philosophy & Technology* 29, no. 4: 307–312. https://doi.org/10.1007/s13347-016-0220-8.

———. 2017. "Group Privacy: A Defence and an Interpretation." In *Group Privacy: New Challenges of Data Technology*. Philosophical Studies Series 126, edited by Linnet Taylor, Luciano Floridi, and Bart van der Sloot, 83–100. Cham: Springer International Publishing.

Floridi, Luciano, and Mariarosaria Taddeo. 2016. "What Is Data Ethics?" *Philosophical Transactions of the Royal Society A: Mathematical, Physical and Engineering Sciences* 374, no. 2083: 1–4. https://doi.org/10.1098/rsta.2016.0360.

Flowerdew, John, and John E. Richardson, eds. 2018. *The Routledge Handbook of Critical Discourse Studies*. Abingdon and New York: Routledge.

Forgó, Nikolaus, Stefanie Hänold, and Benjamin Schütze. 2017. "The Principle of Purpose Limitation and Big Data." In *New Technology, Big Data and the Law*, edited by Marcelo Corrales, Mark Fenwick, and Nikolaus Forgó, 17–42. Singapore: Springer Nature. https://doi.org/10.1007/978-981-10-5038-1_2.

Freude, Alvar C. H., and Trixy Freude. 2016. "Echoes of History: Understanding German Data Protection." *Newpolitik*, 1 October. Washington, DC: Bertelsmann Foundation. http://www.bfna.org/research/echos-of-history-understanding-german-data-protection.

Froud, David. 2018. "GDPR: It's Not Just about EU Citizens, or Residents." Froud on Fraud Blog, 6 February. http://www.davidfroud.com/gdpr-not-just-eu-citizens-or-residents.

GDPR EU.org. n.d. "Fines and Penalties." https://www.gdpreu.org/compliance/fines-and-penalties.

GDPR.Report. 2017. "GDPR: Guidelines and Consequences for Non-compliance." 16 June. https://gdpr.report/news/2017/06/16/gdpr-guidelines-consequences-non-compliance.

General Data Protection Regulation (GDPR). 2016. "Regulation (EU) 2016/679." *Official Journal of the European Union* (4 May): 1–68. https://eur-lex.europa.eu.

GOV.UK. 1998. "Data Protection Act (DPA)." http://www.legislation.gov.uk/ukpga/1998/29/contents.

Hijmans, Hielke, and Charles Raab. 2018. "Ethical Dimensions of the GDPR." In *Commentary on the General Data Protection Regulation*, edited by Mark Cole and Franziska Boehm, 1–14. Cheltenham: Edward Elgar. https://papers.ssrn.com/sol3/papers.cfm?abstract_id=3222677.

ICO (Information Commissioner's Office). 2017. *Big Data, Artificial Intelligence, Machine Learning and Data Protection* (Version 2.2). 4 September. https://ico.org.uk/for-organisations/guide-to-data-protection/big-data.

———. 2018a. *The Guide to Data Protection* (Version 2.10.2). 8 February. https://ico.org.uk/for-organisations/guide-to-data-protection/big-data.

———. 2018b. *Guide to the General Data Protection Regulation* (Version 1.0.17). 22 March. https://ico.org.uk/for-organisations/guide-to-the-general-data-protection-regulation-gdpr.

Kitchin, Rob. 2014a. "Big Data, New Epistemologies and Paradigm Shifts." *Big Data & Society* 1, no. 1: 1–12. https://doi.org/10.1177/2053951714528481.

———. 2014b. *The Data Revolution: Big Data, Open Data, Data Infrastructures and Their Consequences*. London: Sage.

Kitchin, Rob, and Gavin McArdle. 2016. "What Makes Big Data, Big Data? Exploring the Ontological Characteristics of 26 Datasets." *Big Data & Society* 3, no. 1: 1–10. https://doi.org/10.1177/2053951716631130.

Lazarus, David. 2018. "Privacy Rights Remain Just an Illusion." *The Los Angeles Times*, 5 January. http://www.latimes.com.

Levy, Karen E. C., and David Merritt Johns. 2016 "When Open Data Is a Trojan Horse: The Weaponization of Transparency in Science and Governance." *Big Data & Society* 3, no. 1: 1–6. https://doi.org/10.1177/2053951715621568.

Madge, Robert. 2018. "GDPR's Global Scope: The Long Story." *MyData Journal*, 12 May. https://medium.com/mydata/does-the-gdpr-apply-in-the-us-c670702faf7f.

Marwick, Alice E., and danah boyd. 2011. "I Tweet Honestly, I Tweet Passionately: Twitter Users, Context Collapse, and the Imagined Audience." *New Media & Society* 13, no. 1: 114–133. https://doi.org/10.1177/1461444810365313.

Mazzocchi, Fulvio. 2015. "Could Big Data Be the End of Theory in Science? A Few Remarks on the Epistemology of Data-Driven Science." *Science & Society* 16, no. 10: 1250–1255. https://doi.org/10.15252/embr.201541001.

McDermott, Yvonne. 2017. "Conceptualising the Right to Data Protection in an Era of Big Data." *Big Data & Society* 4, no. 1: 1–7. https://doi.org/10.1177/2053951716686994.

Metcalf, Jacob, and Kate Crawford. 2016. "Where Are Human Subjects in Big Data Research? The Emerging Ethics Divide." *Big Data & Society* 3, no. 1: 1–14. https://doi.org/10.1177/2053951716650211.

Metcalf, Jacob, Emily F. Keller, and danah boyd. 2016. "Perspectives on Big Data, Ethics, and Society." *Council for Big Data, Ethics, and Society* (7 July): 1–23. http://bdes.datasociety.net/council-output/perspectives-on-big-data-ethics-and-society.

Meyer, David. 2018. "Activists Are Already Targeting Google and Facebook Over Europe's New Data Privacy Law That Went Live Today." *Fortune*, 25 May. http://fortune.com/2018/05/25/google-facebook-gdpr-forced-consent.

Mittelstadt, Brent Daniel, Patrick Allo, Mariarosa Taddeo, Sandra Wachter, and Luciano Floridi. 2016. "The Ethics of Algorithms: Mapping the Debate." *Big Data & Society* 3, no. 2: 1–21. https://doi.org/10.1177/2053951716679679.

Puaschunder, Julia M. 2017. "The Nudging Divide in the Digital Big Data Era." *International Journal of Research in Business, Economics and Management* 4: 11–12, 49–53. https://ssrn.com/abstract=3007085.

Puaschunder, Julia M. 2018. "Nudgital: Critique of Behavioral Political Economy." *Proceedings of the 9th International RAIS Conference on Social Sciences and Humanities*, 87–117. http://dx.doi.org/10.2139/ssrn.3179017.

Salmons, Janet. 2017a. "Getting to Yes: Informed Consent in Qualitative Social Media Research." In *The Ethics of Online Research (Advances in Research Ethics and Integrity, Volume 2)*, edited by Kandy Woodfield, 109–134. Bingley: Emerald Publishing Limited.

———. 2017b. "What Does 'Informed Consent' Mean Today?" *MethodSpace*, 21 March. https://www.methodspace.com/informed-consent-mean-today.

Schadt, Eric E. 2012. "The Changing Privacy Landscape in the Era of Big Data." *Molecular Systems Biology* 8, no. 1, article number 612. https://doi.org/10.1038/msb.2012.47.

Selbst, Andrew D., and Julia Powles. 2017. "Meaningful Information and the Right to Explanation." *International Data Privacy Law* 7, no. 4: 233–242. https://doi.org/10.1093/idpl/ipx022.

Tannen, Deborah, Heidi E. Hamilton, and Deborah Schriffin, eds. 2015. *The Handbook of Discourse Analysis*, Second Edition. Oxford: Wiley Blackwell.

Taylor, Linnet, Luciano Floridi, and Bart van der Sloot. 2017. *Group Privacy: New Challenges of Data Technology*. Philosophical Studies Series 126. Cham: Springer International Publishing.

van Deursen, Alexander J. A. M., and Karen Mossberger. 2018. "Any Thing for Anyone? A New Digital Divide in Internet-of-Things Skills." *Policy & the Internet* 54: 1–19. https://doi.org/10.1002/poi3.171.

van Dijck, José. 2014. "Datafication, Dataism and Dataveillance: Big Data between Scientific Paradigm and Ideology." *Surveillance & Society* 12, no. 2: 187–208. https://doi.org/10.24908/ss.v12i2.4776.

Veale, Michael, and Reuben Binns. 2017. "Fairer Machine Learning in the Real World: Mitigating Discrimination without Collecting Sensitive Data." *Big Data & Society* 4, no. 2: 1–17. https://doi.org/10.1177/2053951717743530.

Wachter, Sandra, Brent Mittelstadt, and Luciano Floridi. 2017. "Why a Right to Explanation of Automated Decision-Making Does Not Exist in the General Data Protection Regulation." *International Data Privacy Law* (20 September): 1–47. http://dx.doi.org/10.2139/ssrn.2903469.

Wang, Amy B. 2017. "A Lawyer Rewrote Instagram's Terms of Use 'in Plain English' So Kids Would Know Their Privacy Rights." *Washington Post*, 8 January. https://www.washingtonpost.com/news/parenting.

Zödi, Zsolt. 2017. "Law and Legal Science in the Age of Big Data." *Intersections: East European Journal of Society and Politics* 3, no. 2: 69–87. https://doi.org/10.17356/ieejsp.v3i2.324.

Zwitter, Andrej. 2014. "Big Data Ethics." *Big Data & Society* 1, no. 2: 1–6. https://doi.org/10.1177/2053951714559253.

CHAPTER 5

Conclusion

Abstract The chapter summarises the critical overview of the big data debate in global news, health and policy discourse and across the public, corporate and academic settings. It illustrates how big data impinges on the entire spectrum of disciplines, from the hard and life sciences to the social sciences and humanities, posing unexplored ethical and methodological issues. The main question is how to accommodate correlation with causation in big data analytics. It argues that the use of big data does not imply the demise of theory, the death of politics or the disappearance of the human subject to whom data custody and stewardship responsibilities are ultimately assigned. Rather, a novel data ethics is seeing the light and with it, a new way of speaking with/of technology.

Keywords Big data · Data activism · Ethics · Social sciences

This research has attempted to frame the big data debate by investigating the linguistic and discursive features of the narratives that this socio-technical phenomenon has generated since the fuzzy emergence of the term within public and popular discourse around the mid-nineties. In particular, the priority given to language studies has intended to provide novel insights into the ways in which the big data debate unfolds in three crucial domains, i.e. the global news media, healthcare, and legal discourse about data protection in the European Union, after the

M. C. Paganoni, *Framing Big Data*,
https://doi.org/10.1007/978-3-030-16788-2_5

General Data Protection Regulation (GDPR) entered into force on 25 May 2018, with reverberations in public discourse and policy making.

Big data coverage in the global news media has highlighted the linguistic and cultural keywords around which this technological storytelling revolves, registering a major disruption after Cambridge Analytica resulting in the resurfacing of ethics in discourse. Extending the investigation to IoT healthcare has shown that the advancements made possible by new technologies and big data analytics seem to justify the use and repurposing of vast amounts of data in the public interest, navigating the straits between privacy violation and more flexible forms of consent that research ethics is trying to devise. Among other things, the choice has highlighted the existence of an interdiscursive overlap across societal domains, and the limited extent to which ordinary people's concerns manage to reach public forums. The GDPR is the first systematic attempt to address possible scenarios generated by the dematerialisation of the data-driven economy (Lycett 2013) in order to prevent ethical breaches that are unique to the big data ecosystem.

The application of a linguistic and discursive approach to the realm of big data has been privileged as relatively unexplored so far, apart from a few investigations on metaphors (Lupton 2014; Puschmann and Burgess 2014; van den Boomen 2014). Doubtless, metaphors make a number of vivid semantic associations to different fields, such as the natural world (*cloud*), especially with liquid material that can be a resource (like *oil*) or a threat (like a *deluge*), the economy (*currency*) and the body (*footprints*, *shadow*), and are a powerful way to address a new concept and engage the collective imagination (Watson 2014).

At the same time, figurative language is more illuminating when it is placed and discussed within a wider discursive context, where metaphors are interpreted as dynamic rather than static constructs, thus avoiding their reification. A case in point is the widespread analogy of big data as the new oil fuelling the economy that is dominant in popularising science and technology communication. The analogy, introduced here in Chapter 2, is premised on the fallacious assumption that, once this immaterial resource has been used, or rather "burnt", then "its usefulness has been depleted and disposing of it" (Marr 2018, n.p.) while, in actual fact, the crucial point with big data sets is exactly the opposite, i.e. that they can be repurposed. The natural—as well as industrial—metaphor of data as the new oil is exposed as inadequate to fully describe the new setting, technological forecasting and political implications of what has been termed "Industry 4.0" (Marr 2016; Reischauer 2018). In particular, comparing data to a natural phenomenon masks all the human activities that are involved in data collection, processing and storage, and may lead to "large-scale ethical breaches" (Hwang and Levy 2015, n.p.). It is important, instead, to constantly relate big data to the socio-technical systems that both create it and use it, espousing a more encompassing view of technology (Watson 2016).

Several issues that will affect the use of big data are at play for public goals in contemporary societies, research and development in the corporate world, and knowledge production and dissemination in academia. In particular, the ethical implications of digital technology should be prioritised. Arguably, the Cambridge Analytica scandal managed to give momentum to the emergence of ethical reflections with regard to big data in public discourse and not just within expert communities. This violation, whereby big data was "weaponised" to influence voters in the 2016 US presidential election, confirmed the worst suspicions about the nullification of democracy, setting in motion the current public debate on data protection. This happens while national and international legal frameworks—especially the GDPR in the European Union—are striving to address the issue of individual privacy rights, corporate compliance and, ultimately, the digital divide and its unequal distribution of knowledge, access and power (van Deursen and Mossberger 2018).

This effort to tackle the impact of big data is also impellent in academia, where it involves not only the hard and life sciences, but also the social sciences (Fuchs 2017) and the humanities (Borgman 2015), leading to the reshaping of traditional analytical models. Central to the debate in the latter is the need to bring back context as essential in big data epistemologies and in research ethics, deconstructing the myth that data can speak for itself.

Sociology in particular, which used to rely on theoretical models and "small" data collected through interviews, focus groups and surveys, is now challenged to rethink theory through the adoption of big data. Without espousing the demise of theory which is even more necessary the greater the data sets are, a possible answer to combine correlation with causation is offered by what has been defined as "symphonic social science", whereby causality is established through the elaboration and explication of multiple empirical examples, like repeat refrains in classical music (Halford and Savage 2017).

Despite recent scandals and the grey areas yet to be clarified, the current technological debate that this research has illustrated resonates with the promise that big data will ultimately be "put to good use" (Lupton 2014, p. 107). The emerging of big data ethics is indeed an opportunity to mobilise a rigid view of technology as intimately disempowering and at the service of the "nudgital society" (Puaschunder 2017, 2018). In recent years there have been several instances of using big data in clear and inclusive ways, aiming at a more equitable distribution of social and political goods. It is the case of activists monitoring environmental risk (Mah 2017) and of the Datactive project—"the politics of data according to civil society"—funded by the European Research Council, which takes "a critical approach to massive data collection. It emerges out of existing activism sub-cultures, such as the hacker and the open-source movements, but overcomes their elitist character to involve also ordinary users" (Datactive, n.d.). It is also the case of the use of big data in mapping health risks, but also of the antitrust watch in order to curb the power and the control over data of companies like Amazon, Facebook, and Google. Another crucial field is the international arena in which big data may facilitate problem-solving and peacekeeping and prevent or quickly respond to "disaster, conflict, health and environmental problems" (Chandler 2015, p. 834).

To conclude, the critical overview of the big data debate at the nodal points of global news, healthcare and policy and across public, corporate and academic settings has shown that its use does not imply the demise of theory, the death of politics or the disappearance of the human subject to whom data custody and stewardship responsibilities are ultimately assigned. Rather, a novel data ethics is seeing the light and with it a new way of speaking with/of technology.

References

Borgman, Christine L. 2015. *Big Data, Little Data, No Data: Scholarship in the Networked World*. Cambridge, MA and London: The MIT Press.
Chandler, David. 2015. "A World without Causation: Big Data and the Coming of Age of Posthumanism." *Millennium: Journal of International Studies* 43, no. 3: 833–851. https://doi.org/10.1177/0305829815576817.
Datactivism. "About." https://data-activism.net.
Fuchs, Christian. 2017. "From Digital Positivism and Administrative Big Data Analytics towards Critical Digital and Social Media Research!" *European*

Journal of Communication 32, no. 1: 37–49. https://doi.org/10.1177/0267323116682804.

Halford, Susan, and Mike Savage. 2017. "Speaking Sociologically with Big Data: Symphonic Social Science and the Future for Big Data Research." *Sociology* 51, no. 6: 1132–1148. https://doi.org/10.1177/0038038517698639.

Hwang, Tim, and Karen Levy. 2015. "'The Cloud' and Other Dangerous Metaphors: Contemporary Ideas about Data and Privacy Are Tied Up Inextricably with Language Choices." *Atlantic*, 20 January. https://www.theatlantic.com/technology/archive/2015/01/the-cloud-and-other-dangerous-metaphors/384518.

Lupton, Deborah. 2014. *Digital Sociology*. Abingdon and New York: Routledge.

Lycett, Mark. 2013. "'Datafication': Making Sense of (Big) Data in a Complex World." *European Journal of Information Systems* 22, no. 4: 381–386. https://doi.org/10.1057/ejis.2013.10.

Mah, Alice. 2017. "Environmental Justice in the Age of Big Data: Challenging Toxic Blind Spots of Voice, Speed, and Expertise." *Environmental Sociology* 3, no. 2: 122–133. https://doi.org/10.1080/23251042.2016.1220849.

Marr, Bernard. 2016. "What Everyone Must Know about Industry 4.0." *Forbes*, 20 June. https://www.forbes.com/sites/bernardmarr/2016/06/20/what-everyone-must-know-about-industry4.0.

———. 2018. "Here's Why Data Is Not the New Oil." *Forbes*, 5 March. https://www.forbes.com/sites/bernardmarr/2018/03/05/heres-why-data-is-not-the-new-oil.

Milan, Stefania. 2017. "Data Activism as the New Frontier of Media Activism." In *Media Activism in the Digital Age*, edited by Viktor Pickard and Goubin Yang, 151–163. Abingdon and New York: Routledge.

Puaschunder, Julia M. 2017. "The Nudging Divide in the Digital Big Data Era." *International Journal of Research in Business, Economics and Management* 4: 11–12, 49–53. https://ssrn.com/abstract=3007085.

Puaschunder, Julia M. 2018. "Nudgital: Critique of Behavioral Political Economy." *Proceedings of the 9th International RAIS Conference on Social Sciences and Humanities*, 87–117. http://dx.doi.org/10.2139/ssrn.3179017.

Puschmann, Cornelius, and Jean Burgess. 2014. "Big Data, Big Questions| Metaphors of Big Data." *International Journal of Communication* 8: 1690–1709. http://ijoc.org/index.php/ijoc/article/view/2169.

Reischauer, Georg. 2018. "Industry 4.0 as Policy-Driven Discourse to Institutionalize Innovation Systems in Manufacturing." *Technology Forecasting and Social Change* 132, no. C: 27–33. https://doi.org/10.1016/j.techfore.2018.02.012.

van den Boomen, Marianne. 2014. *Transcoding the Digital: How Metaphors Matter in New Media*. Amsterdam: Institute of Network Cultures.

van Deursen, Alexander J. A. M., and Karen Mossberger. 2018. "Any Thing for Anyone? A New Digital Divide in Internet-of-Things Skills." *Policy & Internet* 10, no. 2: 122–140. https://onlinelibrary.wiley.com/doi/full/10.1002/poi3.171.

Watson, Sara M. 2014. "Data Is the New '____': Sara M. Watson on the Industrial Metaphor of Big Data." *DIS*. http://dismagazine.com/discussion/73298/sara-m-watson-metaphors-of-big-data.

———. 2016. "Toward a Constructive Technology Criticism." *Tow Center for Digital Journalism White Papers*. New York: Columbia University. https://doi.org/10.7916/D86401Z7.

Index

© The Editor(s) (if applicable) and The Author(s), under exclusive licence to Springer Nature Switzerland AG 2019
M. C. Paganoni, *Framing Big Data*,
https://doi.org/10.1007/978-3-030-16788-2